Heinrich Hofmann
Zakia Rahman
Ulrich Schubert
(eds.)

Nanostructured Materials

SpringerWienNewYork

Prof. Dr. Ing. Heinrich Hofmann
Département des Matériaux, École Polytechnique Fédéral Lausanne,
Lausanne, Switzerland

Dr. Zakia Rahman
Department of Physics, Magnetics Research Laboratory, University of Limerick,
Limerick, Ireland

Prof. Dr. Ulrich Schubert
Institute of Materials Chemistry, Vienna University of Technology,
Vienna, Austria

Softcover reprint of the hardcover 1st edition 2002

Cover illustration:
L. Soriano et al., „Electronic Structure ...", Fig. 1, page 115, this volume
Typesetting: Thomson Press (India) Ltd., New Delhi

Printed on acid-free and chlorine-free bleached paper
SPIN: 10863947
With numerous Figures and Tables
CIP data applied for

Special Edition of
Monatshefte für Chemie/Chemical Monthly, Vol. 133, No. 6, 2002

ISBN-13: 978-3-7091-7398-5 e-ISBN-13: 978-3-7091-6740-3
DOI: 10.1007/978-3-7091-6740-3

Editorial

Nanotechnology can be defined as the development, fabrication, and utilization of materials, devices, and systems through the control of the structure on the nanometer scale, that is, at the level of atoms, molecules, crystals, and supramolecular structures. The challenge for researchers working in the field of nanotechnology is the ability to work at these levels to generate larger structures with fundamentally new atomic, molecular, or particle organization. These nanostructures, assembled from building blocks understood from first principles, exhibit novel physical, chemical, mechanical, and biological properties and phenomena. Control of materials on the nanoscale already plays an important role in scientific disciplines as diverse as physics, chemistry, materials science, biology, medicine, engineering, and computer simulation. For example, it has been shown that carbon nanotubes have a 60 times higher specific strength than steel, and that nanoparticles can target and kill cancer cells. All natural materials and systems establish their foundation at the nanoscale; control of matter at molecular levels means tailoring properties, phenomena, and processes exactly at the scale where the basic properties are determined. Nanotechnology, and especially nanostructured materials as well as materials for nanotechnology will be a strategic branch of science and engineering for the next century. This was the reason why the COST action "Nanostructured Materials" was created in 1998. This action, one of the largest in the field of material science, is focused on fundamental as well as application related aspects of the development of nanostructured materials including modelling and simulation of the properties of such materials. All aspects of materials, mechanical, magnetic, optical, chemical, and electric properties, and the synthesis and characterization of such structures are included in this large research program in which 24 European countries are participating.

This special issue of *Chemical Monthly* gives an overview of the abovementioned activities of the COST action on the occasion of the mid term meeting of the action in Limerick, October 3–4, 2001. It contains a selection of talks given there which reflect the broad and interesting research area of nanostructured materials as well as the high standard of the participating research groups. The editors hope that this collection stimulates the interest of the materials science community to reinforce research activities in the field of application of such novel materials. This means that, beside the fundamental research regarding the properties of nanosized building blocks, new and economically interesting fabrication methods of bulk material as well as of nanostructured materials and nanosized structures which are synthesised directly on devices or chips have to be developed.

Heinrich Hofmann
(*Chairman of COST 523*)
Zakia Rahman
Ulrich Schubert

Contents

Abstracted/Indexed in:
Current Contents, SCI, ASCA, and Current Abstracts of Chemistry,
Chemical Abstracts, Polymer Contents, ADONIS

Invited Review

Magnetic Nanoparticles and Biosciences

Ivo Šafařík[1,2,*] and **Mirka Šafaříková**[1]

[1] Laboratory of Biochemistry and Microbiology, Institute of Landscape Ecology, CZ-37005 České Budějovice, Czech Republic
[2] Department of General Biology, University of South Bohemia, CZ-37005 České Budějovice, Czech Republic

Summary. Magnetic nanoparticles represent an interesting material both present in various living organisms and usable for a variety of bioapplications. This review paper will summarize the information about biogenic magnetic nanoparticles, the ways to synthesize biocompatible magnetic nanoparticles and complexes containing them, and the applications of magnetic nanoparticles in various areas of biosciences and biotechnologies.

Keywords. Magnetic nanoparticles; Biosciences; Magnetic properties; Nanostructures.

Introduction

Nanotechnology involves the study, control, and manipulation of materials at the nanoscale, typically having dimensions up to 100 nm. This is a truly multidisciplinary area of research and development. The large interest in nanostructures results from their numerous potential applications in various areas such as materials and biomedical sciences, electronics, optics, magnetism, energy storage, and electrochemistry [1].

Magnetic nanoparticles having connections to biological systems and bioapplications usually exist or can be prepared in the form of either single domain or superparamagnetic magnetite (Fe_3O_4), greigite (Fe_3S_4), maghemite (γ-Fe_2O_3), various types of ferrites ($MeO \cdot Fe_2O_3$, where $Me = $ Ni, Co, Mg, Zn, Mn, ...), iron, nickel *etc.* Synthetic magnetic nanomaterials are most often available in the form of magnetic fluids (ferrofluids) which have already found many interesting applications (sealing, damping, heat transfer, levitation of magnetic and nonmagnetic objects, production of computers, loudspeakers, measuring devices *etc.*) [2]. Potential applications of biocompatible magnetic fluids in various areas of biosciences and biotechnologies are also very promising as will be shown in this paper.

* Corresponding author. E-mail: safarik@uek.cas.cz

The first magnetic nanoparticles necessary to create a stable magnetic fluid were prepared in the early 1960s [3]. On the contrary, various organisms living on Earth have been able to synthesize them for a long time. These biogenic magnetic nanoparticles have in many respects even better properties than nanoparticles prepared in the laboratory.

The purpose of this review article is to show the close connection between inorganic magnetic nanoparticles and living systems.

Biogenic Magnetic Nanoparticles

In 1962, *Lowenstam* first discovered biochemically precipitated magnetite as a capping material in the radula (tongue plate) teeth of chitons (marine mollusks of the class Polyplacophora) [4]. *Lowenstam* was able to demonstrate the biological origin of this material through a variety of radioisotope tracing studies and by detailed examination of the tooth ultrastructure. Prior to this discovery, magnetite was thought to form only in igneous or metamorphic rocks under high temperatures and pressures. In the chitons, the magnetite serves to harden the tooth caps, enabling the chitons to extract and eat endolithic algae from within the outer few millimetres of rock substrates. Metabolic iron is at first transported to the posterior end of the radula sac and is then deposited as the mineral ferrihydrite within a preformed proteinaceous mesh, forming one or two distinct rows of reddish teeth. This ferrihydrite is converted rapidly to magnetite *via* an unknown process [5].

In 1975, *Blakemore* discovered magnetotactic bacteria [6], which now represent the most intensively studied biomagnetic system. Magnetotactic bacteria form a heterogeneous group of *Gram*-negative prokaryotes with morphological and habitat diversity which have an ability to synthesize fine (50–100 nm) intracellular membrane-bound ferromagnetic crystalline particles (Fig. 1) consisting of magnetite (Fe_3O_4) or greigite (Fe_3S_4) which are covered with an intracellular phospholipid membrane vacuole, forming structures called 'magnetosomes' [7]. Various morphological types of magnetotactic bacteria such as cocci, short or long rods, vibrios, spirilla (Fig. 2), and multicellular forms have been isolated from sediments in diverse aquatic environments, *e.g.*, marine, river, lake, pond, beach, rice paddies, drains, wet soil, deep see, and estuary [8]. Chains of magnetosomes act as simple compass needles which passively torque the bacterial cells into alignment with the earth's magnetic field and allow them to seek the microaerophilic zone at the mud/water interface of most natural aqueous environments. These bacteria swim to the magnetic north in the northern hemisphere, to the magnetic south in the southern hemisphere, and both ways on the geomagnetic equator [5]. Excellent up-to-date reviews covering all aspects of the topic exist [7–9].

Extracellular production of nanometer magnetite particles by various types of bacteria has also been described [10]. Magnetite-bearing magnetosomes have been found in eukaryotic magnetotactic algae, each cell containing several thousand crystals [11]. Biogenic nanometer magnetite particles have been found in marine and lake sediments [12].

The behaviour of various other organisms is also influenced by changes of the magnetic field. It has been shown that some animals, including ants [13],

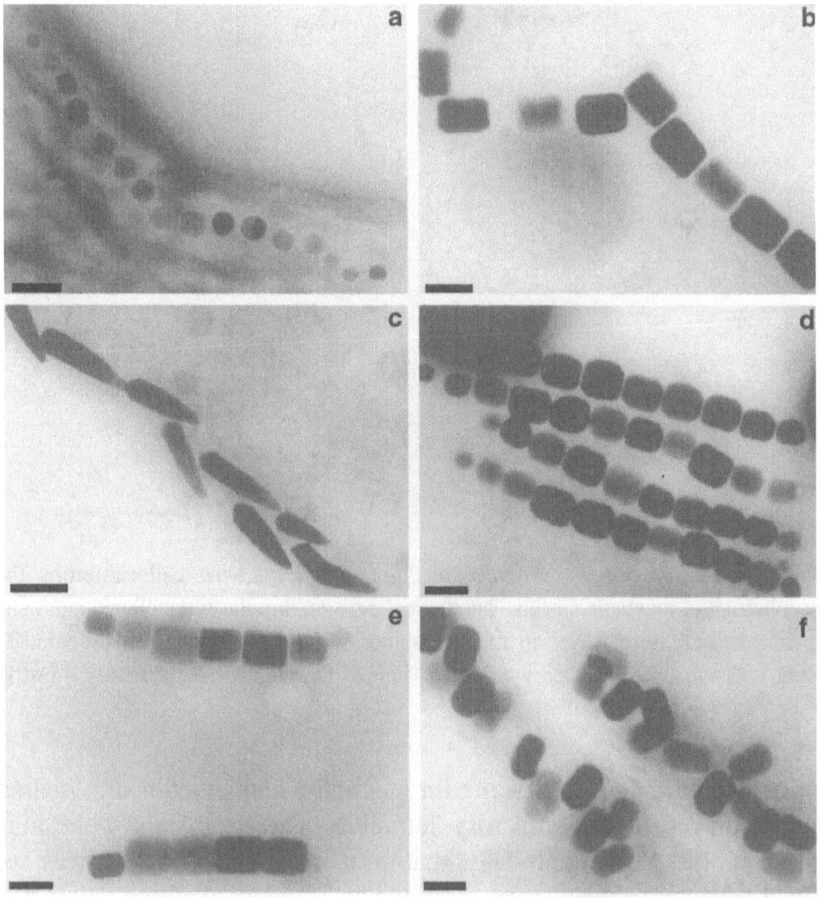

Fig. 1. Electron micrographs of crystal morphologies and intracellular organization of magnetosomes found in various magnetotactic bacteria. Shapes of magnetic crystals include cubo-octahedral (a), elongated hexagonal prismatic (b, d, e, f), and bullet-shaped morphologies (c). The particles are arranged in one (a, b, c), two (e), or multiple chains (d) or irregularly (f). The bar is equivalent to 100 nm. Reproduced with permission of Dr. *D. Schüler*, Germany, from Ref. [7]

honeybees [14], homing pigeons [15], salmon [16], and others use geomagnetic field information for orientation, homing, and foraging. Several hypotheses have appeared trying to explain the mechanisms of magnetoreception [17]. It seems apparent that biomineralized magnetite nanoparticles can interact with the geomagnetic field, monitoring information on its intensity and direction and being thus the main part of a highly evolved, finely tuned sensory system [18]. Identifying the presence of magnetite particles in different organisms whose behaviour is influenced by the geomagnetic field is a first step towards demonstrating that biogenic magnetite is involved in geomagnetic field detection [19]. Table 1 shows some typical examples of living organisms containing magnetic nanoparticles within their cells and organs.

Rainbow trout (*Oncorhynchus mykiss*) has been used for extensive study of the magnetoreception mechanism. The key behavioural, physiological, and anatomical components of a magnetite-based magnetic sense have been demonstrated.

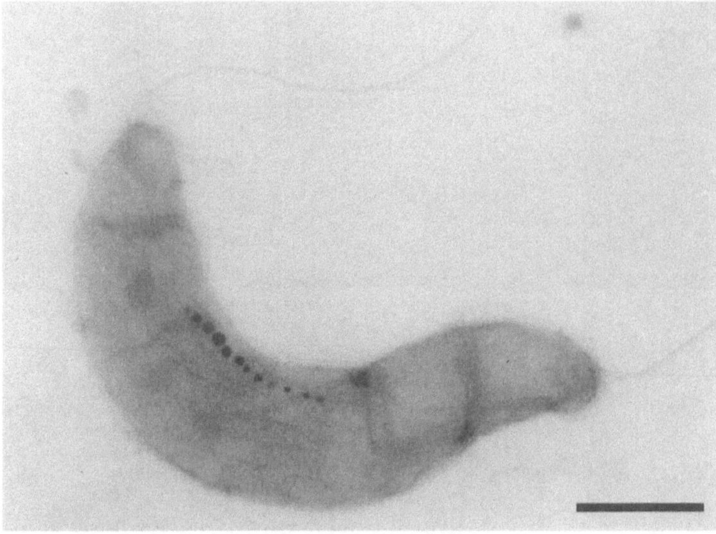

Fig. 2. Electron micrograph of a *Magnetospirillum gryphiswaldense* cell exhibiting the characteristic morphology of magnetic spirilla. The helical cells are bipolarly flagellated and contain up to 60 intracellular magnetite particles in magnetosomes which are arranged in a chain. The bar is equivalent to 0.5 μm. Reproduced with permission of Dr. *D. Schüler*, Germany, from Ref. [7]

Receptor cells containing single-domain magnetite nanocrystals are located within a discrete sublayer of the olfactory lamellae. These crystals were mapped to individual receptors using confocal and atomic force microscopy. It was confirmed that several magnetic crystals are arranged in a chain of about 1 μm within the receptor, and that the receptor is a multi-lobed single cell. These results are consistent with a magnetite-based detection mechanism, as 1 μm-chains of single-domain magnetite crystals are highly suitable for the behavioural and physiological responses to magnetic intensity [20, 21].

The information given so far is connected with the physiological occurrence of magnetic nanoparticles in given tissues and organs. However, it has been shown that many neurodegenerative diseases are connected with the disruption of normal iron homeostasis in the brain. Nanoscale magnetic biominerals (primarily magnetite and maghemite) may be associated with senile plaques and tau filaments found in brain tissue affected by these diseases. These findings have important implications for our understanding of the role of iron in neurodegenerative diseases as well as profound implications for their causes. In addition, the presence of biogenic magnetite in affected tissue should also provide improved mechanisms for early detection through the modification of MRI pulse sequences [22].

Recently, an interesting observation has been presented by NASA researchers. During the detailed study of Martian meteorites, magnetite nanoparticles were found which were similar to those present in magnetotactic bacteria (Fig. 3). *Thomas-Keprta et al.* [23] postulated six criteria that characterize biologically produced magnetite crystals. The simultaneous presence of all six characteristics — *i.e.*, a definite size range and width/length ratio, chemical purity, crystallographic perfection, arrangement of crystals in linear chains, unusual crystal morphology, and

Table 1. Examples of organisms synthesizing magnetic nanoparticles (MNP)

Type of organism	General name	Latin name	Localisation of MNP	Type of MNP	References
Micro-organisms	Magnetotactic bacteria	*Magnetospirillum sp.*	magnetosomes	Fe_3O_4	[8]
			magnetosomes	Fe_3S_4	[149]
	Algae		cells	Fe_3O_4	[11]
Protozoa			cells	Fe_3O_4	[150]
Insect	Honeybee	*Apis mellifera*	abdomen	Fe_3O_4	[14]
	Migratory ant	*Pachycondyla marginata*	abdomen	Fe_3O_4	[13]
	Termites	*Nasutitermes exitiosus*	thorax, abdomen	Fe_3O_4	[151]
		Amitermes meridionalis	thorax, abdomen	Fe_3O_4	[151]
Fish	Atlantic salmon	*Salmo salar*	lateral line	Fe_3O_4	[152]
	Sockeye salmon	*Oncorhynchus nerka*	skull	Fe_3O_4	[153]
	Rainbow trout	*Oncorhynchus mykiss*	olfactory lamellae	Fe_3O_4	[21]
	Chum salmon	*Oncorhynchus keta*	head	Fe_3O_4	[16]
Amphibians	Eastern red-spotted newt	*Notophthalmus viridescens*	whole body	Fe_3O_4	[154]
Birds	Bobolink	*Dolichonyx oryzivorus*	upper beak	Fe_3O_4	[155, 156]
	Homing pigeon	*Columba livia*	upper-beak skin	Fe_3O_4	[15, 157]
Mammals	Common Pacific dolphin	*Delphinus delphis*	dura mater	Fe_3O_4	[158]
	Human	*Homo sapiens*	brain, heart	Fe_3O_4	[159–163]

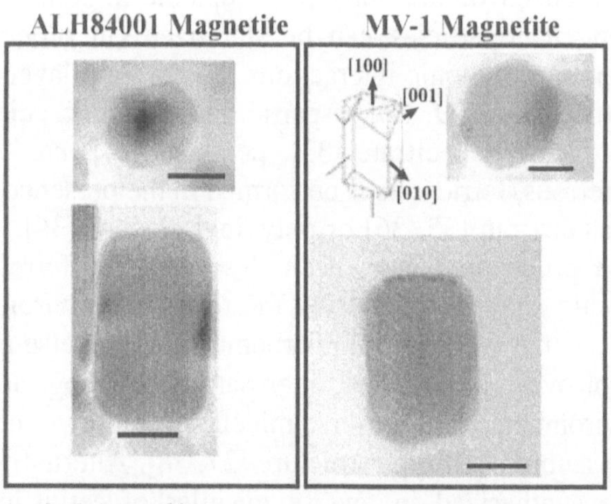

Fig. 3. Typical examples of magnetic nanoparticles found in meteorite ALH84001 (originating from Mars) and in magnetotactic bacterium MV-1. Reproduced with permission of Dr. *K. L. Thomas-Keprta*, USA

elongation of crystals in the [111] crystallographic direction — should constitute evidence of biological origin. During the research, magnetic nanoparticles present in meteorite ALH84001 fulfilled all six criteria. These findings led to the hypothesis that they could be in fact microfossils of former Martian magnetotactic bacteria [23, 24]. Subsequently another hypothesis has appeared suggesting life on Earth could originally have arrived here by way of meteorites from Mars, where conditions early in the history of the solar system are thought to have been more favourable for the creation of life from nonliving ingredients. It was calculated that under the optimal conditions the temperature within the meteorite did not exceed 40°C; the transport time from Mars to Earth could take only a few years, and thus meteorites could transfer life between planets [25].

Biocompatible Magnetic Nanoparticles and Complexes Containing Them

Various types of magnetic fluids (ferrofluids) are often used as the starting material to prepare the target nanomaterials. Ferrofluids are colloidal solutions of magnetic iron oxide (or ferrite) nanoparticles (around 10 nm in diameter) in either a polar or non-polar liquid. These particles are said to be 'superparamagnetic', meaning that they are attracted by a magnetic field but retain no residual magnetism after the field is removed. The first ferrofluids were prepared by grinding micrometer-sized magnetic particles in a ball mill for several weeks and subsequent transfer of nanometer particles into kerosene. The particles were stabilized with oleic acid to prevent clumping. Nowadays, however, the chemical synthesis of ferrofluids preferably employs coprecipitation of ferric and ferrous salts with alkaline solution and subsequent treatment under hydrothermal conditions. To form ferrites, other divalent ions are used instead of ferrous ions [26].

Biocompatible ferrofluids normally use water as a carrier medium, but paraffin or vegetable oils based ferrofluids may be acceptable in some case. In the water phase, the magnetic nanoparticles can be stabilized (in order to prevent their unwanted agglomeration) by ionic interactions [27, 28], a bilayer of an appropriate surfactant (*e.g.* fatty acids) [29, 30], aspartic and glutamic acid [31], *meso*-2,3-dimercaptosuccinic acid [32], citrate [33], peptides [34], *etc.* Alternatively, the coprecipitation of ferrous/ferric ions is performed in the presence of an appropriate biopolymer, such as dextran [35, 36] or polyvinyl alcohol [36].

Many modified procedures have been described for ferrofluid preparation. Synthesis of magnetic nanoparticles using the restricted environments offered by surfactant systems such as water-in-oil microemulsions (reverse micelles) provides an excellent control over particle size, inter-particle spacing, and particle shape. The controlled environment of the reverse micelle also allows sequential synthesis which can produce a core-shell type structure [37, 38]. Alternatively, metallic iron nanoparticles were synthesized in reverse micelles of cetyltrimethylammonium bromide (*CTAB*) using hydrazine as a reducing agent. After addition of an aqueous gold solution, a metallic gold coating on the outer surface of the iron particles was formed. The gold shells on the iron particles provide functionality with thiol-functionalized substrates [39].

Magnetoliposomes are magnetic derivatives of liposomes and can be prepared by entrapment of ferrofluids within the core of liposomes [40, 41]. Affinity magnetoliposomes can be produced by covalent attachment of ligands to the surface of the vesicles or by incorporation of target lipids in the matrix of structural phospholipids [42]. Alternatively, magnetoliposomes are prepared using the phospholipid vesicles as nanoreactors for the *in situ* precipitation of magnetic nanoparticles [43]. Vesicles of another type are constituted by didodecyldimethylammonium bromide, contain an ionic magnetic fluid, and have a diameter of about 1 μm [44].

Micrometric particles (*e.g.* materials used for column chromatography) have been post-magnetized by circulation of ferrofluid through the column chromatography carrier [45]. Dynabeads (magnetic polystyrene particles of a diameter of 2.8 or 4.5 μm produced by Dynal, Norway) are prepared in magnetic form after introduction of $-NO_2$ or $-ONO_2$ groups to the matrix and subsequent reaction with Fe^{2+} salts. During the reaction, iron hydroxides precipitate inside the pores, and after heating they are transformed into nanoparticles of γ-Fe_2O_3 [46]. Ferrofluids can be added to reaction mixtures used to create various magnetic polymeric microparticles [47, 48]. Ferrite plating enables magnetization of various biocompatible materials by directed precipitation of magnetic iron oxides [49, 50]. Another type of magnetic polymeric particles has been prepared by electrostatic adsorption of negatively charged magnetic nanoparticles on positively charged polymer particles and subsequent encapsulation of the prepared complex [51, 52].

Magnetic nanocomposites have been prepared using cross-linked synthetic polymers and different polysaccharides like alginate or cellulose as gel matrices [53–55]. Another procedure is based on the nucleation and growth control of crystalline iron oxide particles in organic matrix through the reaction control of a

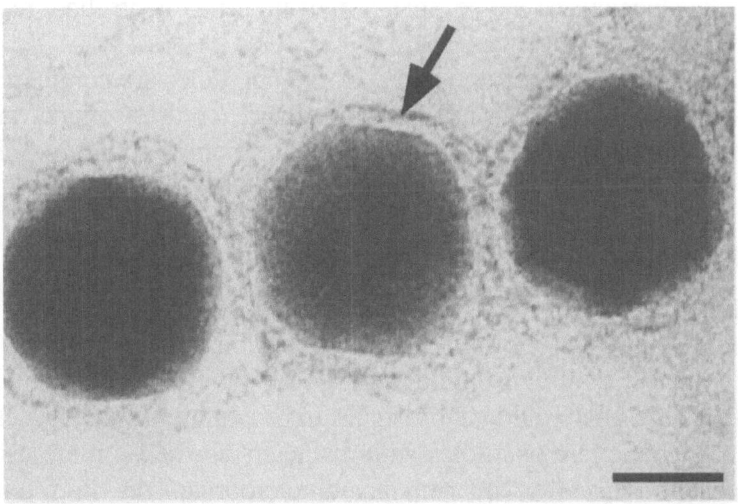

Fig. 4. Magnetosome particles isolated from *M. gryphiswaldense*. The magnetite crystals are typically 42 nm in diameter and are surrounded by the magnetosome membrane (arrow). The bar is equivalent to 25 nm. Reproduced with permission of Dr. *D. Schüler*, Germany, from Ref. [7]

Table 2. Examples of commercially available biocompatible magnetic nanoparticles

Product name	Composition	Particle size (nm)	Application	Manufacturer or supplier
Combidex	Magnetic iron oxides – dextran	17–20	Magnetic resonance contrast agent	Advanced Magnetics, USA
Endorem[a]/ Feridex[b]	Magnetic iron oxides – dextran	100–250	Magnetic resonance contrast agent	Advanced Magnetics, USA
MicroBeads	Magnetic iron oxides – dextran	50	Separation and labelling of cells and molecules	Miltenyi Biotec, Germany
Nanomag	Magnetic iron oxides – dextran	100	Magnetic labelling	Micromod Partikeltechnologie, Germany
Resovist	Magnetic iron oxides – dextran	57	Magnetic resonance contrast agent	Schering AG, Germany

[a] Commercial product name in Europe; [b] commercial product name in the United States

metallorganic precursor with a combination of the hydrolysis and polymerization below 100°C. The reaction conditions influence the size and crystallinity of magnetic nanoparticles in the organic matrix [56].

Bacterial magnetite nanoparticles obtained from magnetotactic bacteria after disruption of the cell wall and subsequent magnetic separation have been used for a variety of bioapplications. Due to the presence of the lipid layer (Fig. 4) the particles are biocompatible, their suspensions are very stable and the particles can be easily modified [8, 57].

Magnetic derivatives of the iron storage protein ferritin (magnetoferritin) have also been prepared. A magnetic mineral has been synthesized within the nanodimensional cavity of horse spleen ferritin using controlled reconstitution conditions [58].

Several types of biocompatible magnetic nanoparticles are commercially available, especially those used as magnetic resonance contrast agents or magnetic labels. A non-complete selection of these products can be found in Table 2.

Application of Magnetic Nanoparticles in Biosciences

Many different types of magnetic micro- and nanoparticles and molecular magnetic labels have been used for a great number of applications in various areas of biosciences and biotechnologies [59–62]. The majority of review papers considers magnetic micro- and nanoparticles to be equally important. Of course, in many areas magnetic microparticles with diameters above 1 μm are used (*e.g.* for immunomagnetic separation of pathogenic microorganisms in food and clinical microbiology), whereas for other applications magnetic nanoparticles are necessary. The following chapters will focus on the bioapplications of magnetic nanoparticles and the most important complex material containing them. Applications of dynabeads (containing magnetic nanoparticles within the bead structure) will not be considered in this review; there are several sources where information about these particles can be found [63, 64].

Immobilization and modification of biologically active compounds

Immobilization of enzymes, antibodies, oligonucleotides, and other biologically active compounds is a very important technique used in various areas of biosciences and biotechnology. Biologically active compounds immobilized on magnetic carriers can be removed from the system by using an external magnetic field or can be targeted to the desired place. The immobilized compounds can be used to express their activities in a desired process (*e.g.* immobilized enzymes) or can be used as affinity ligands enabling to capture or modify the target molecules or cells.

Magnetic nanoparticles obtained from magnetotactic bacteria have been used for the immobilization of a variety of enzymes, such as glucose oxidase and uricase [65], antibodies [66–68], oligonucleotides [69, 70], *etc.* Colloidal aqueous suspensions of superparamagnetic nanoparticles (9 nm in diameter) composed of maghemite and forming an ionic ferrofluid have been covalently coupled with lectins, enzymes, or antibodies using specific thiol chemistry [71, 72]. Magnetic nanoparticles activated with 3-aminopropyltriethoxysilane have been used for the immobilization of various enzymes, antibodies, and protein A after glutaraldehyde treatment [73, 74]. Dextran-based biocompatible magnetic nanoparticles (*ca.* 50 nm in diameter) produced commercially by Miltenyi Biotec, Germany, are available with many covalently immobilized molecules (*e.g.* annexin V, antibiotin antibody, antifibroblast antibody, anti-FITC antibody, anti-HLA-DR antibody, antihuman epithelial antigen antibody, antihuman melanoma-associated chondroitin sulfate proteoglycan antibodies, antimouse DX5 antibody, antiphycoerythrin antibody, antibodies against various CD markers, goat antimouse, antirabbit, and antirat IgGs, various types of rat antimouse IgG and IgM antibodies, protein A, protein G, streptavidin, and some others; see http://www.miltenyibiotec.com).

Enzymes can be made soluble and active in organic solvents by chemical modification with the amphipathic macromolecule polyethylene glycol (*PEG*). The *PEG*-enzyme conjugates can be also conjugated to magnetic nanoparticles. Alternatively, the magnetite-*PEG* conjugate is prepared first and then conjugated to the target enzyme. The magnetically modified enzymes stably disperse in both organic solvents and aqueous solutions. Magnetically modified lipase catalyzes ester synthesis in organic solvents and can be easily recovered by magnetic force without loss of enzyme activity [75, 76]. Other enzymes such as *L*-asparaginase [76] and urokinase [77] have also been modified in this way.

Antibodies can be modified in a similar manner. Magnetite-labelled antibodies are expected to be applicable clinically as a therapeutic agents for the induction of hyperthermia [78].

Magnetoliposomes containing magnetic nanoparticles entrapped within the cavity have been used for the immobilization of membrane-bound enzymes [79] or antibodies [80] as well as for the entrapment of various drugs [81]. The catalytic activity of isolated lipid-depleted membrane-bound enzymes, such as cytochrom c-oxidase, has been substantially enhanced after their incorporation in magneto-liposomes.

Production of a protein (enzyme, antibody, protein A) – magnetite complex by genetically engineered magnetotactic bacteria *Magnetospirillum sp.* AMB-1 has been proposed recently [82]. The *magA* gene, encoding an integral iron translocating protein situated in the membrane of *Magnetospirillum sp.* magnetic nanoparticles, has been fused with genes encoding the desired protein. The desired protein-*magA* fusion gene has been cloned into *Magnetospirillum sp.*, and after cultivation, magnetic nanoparticles bearing the desired protein on the particle membrane surface have been isolated. The *magA* protein may thus be used as an anchor for the site-specific expression of foreign proteins on bacterial magnetite particle membranes by gene fusion, thus obviating the need for immobilization of the proteins using chemical reagents [83]. Bacterial magnetic nanoparticles containing protein A [84], luciferase [85], and acetate kinase [85] have already been constructed.

Isolation of biologically active compounds

The isolation and separation of specific molecules belongs to the major problems in biosciences. Affinity ligand techniques represent currently the most powerful tool available to the downstream processing both with respect to selectivity and recovery. Batch magnetic isolations may be faster than standard liquid chromatography procedures, and the target molecules can be separated from untreated samples containing impurities. Although magnetic microparticles are usually employed for this purpose, especially when working with larger volumes of solutions and suspensions, magnetic nanoparticles have been used to a greater extent recently.

Isolation of eukaryotic poly(A)$^+$m*RNA* can be performed using oligo(d*T*) immobilized on synthetic magnetic nanoparticles [86]. Alternatively, oligonucleotides immobilized on bacterial magnetic particles can be used for the same purpose [70]. Protein A immobilized on magnetic nanoparticles has been used to purify monoclonal antibodies [87]. Alcohol dehydrogenase and lactate dehydrogenase have been isolated using ferrofluid-modified 5′-AMP-sepharose 4B as an affinity adsorbent, whereas ferrofluid-modified 2′,5′-ADP-sepharose 4B was used to isolate glucose-6-phosphate dehydrogenase and 6-phosphogluconate dehydrogenase. IgG and anti-human serum albumin antibodies have been isolated using ferrofluid-modified protein A sepharose and human serum albumin sepharose, respectively [45, 88].

Biocompatible two-phase systems, composed for example from aqueous dextran and polyethylene glycol phases, have been used for the isolation of biologically active compounds, subcellular organelles, and cells for many years. To speed up the phase separation the addition of ferrofluids to the system is useful. In a magnetic field such additives will induce a faster phase separation. Dextran-stabilized ferrofluid added to an aqueous two-phase system containing polyethylene glycol and dextran totally partitioned to the dextran phase. After mixing of the two-phase system, it was possible to reduce the separation time by a factor of 35 by applying a magnetic field to the system [89]. Other types of surface modified ferrofluids have also been successfully used [90].

Determination and detection of biologically active compounds and xenobiotics

An enormous amount of analytical procedures for the detection of biologically active compounds is available; among them, antibody-based techniques are exceptionally important. Magnetic modifications of standard immunoassays can be successfully used for the determination of various biologically active compounds and xenobiotics. In these techniques, specific antibodies or antigens are covalently immobilized on fine magnetic particles. Magnetically based assays are usually faster and more reproducible than the standard microtitration plate based assays and have proven to be simple, rapid, and sensitive. The detection systems can be based on the use of enzymes, radioisotopes, fluorescent substances, or chemiluminescence. The possibility of automation of the analytical procedures is especially important.

Magnetic nanoparticles with immobilized antimouse IgG antibody or protein A have been applied to enzyme-linked immunosorbent assay (ELISA) of mouse IgG. The assay time could be shorten substantially in comparison with the conventional method [91]. Using bacterial magnetic nanoparticles as a carrier, a highly sensitive mouse IgG assay was developed having a good relationship between the luminescence intensity and mouse IgG concentration in the range of $1–10^5$ fg/cm^3 [66]. Antibody-conjugated bacterial magnetic particles have been prepared to determine a model food allergen lysozyme using a high-performance and rapid chemiluminescence immunoassay and a fully automated system [92]. A fully automated sandwich immunoassay for the determination of human insulin using an antibody – protein A – bacterial magnetic particle complexes and an alkaline phosphatase-conjugated secondary antibody has been described recently [83].

Bacterial magnetic nanoparticles have been used for the identification of cyanobacterial *DNA*. Genus-specific oligonucleotide probes for the detection of target strains were designed from the variable region of the cyanobacterial 16S r*DNA*. These oligonucleotide probes were immobilized on magnetic particles *via* streptavidin-biotin conjugation and employed for magnetic capture (hybridization) against digoxigenin-labelled cyanobacterial 16S r*DNA*. Bacterial magnetic particles were magnetically concentrated, spotted in a microarray device, and the fluorescent detection was performed. The cyanobacterial genera were successfully discriminated [69]. A similar approach has been used for the discrimination between Atlantic and Pacific subspecies of the northern bluefin tuna (*Thunnus thynnus*) [93].

Alternative immunoassays or assays employing other binding molecules (*e.g.* lectins) can employ magnetic nanoparticles as ferromagnetic labels instead of enzymes, radionuclides, *etc.* After binding the labelled antibody or lectin to the target analyte, the magnetically labelled complex can be captured by larger particles with immobilized specific antibodies against another analyte epitope, and after sedimentation the amount of magnetically labelled analyte in the sedimented fraction is measured with an appropriate transducer (*e.g.* a magnetic permeability meter). The possible advantage of this approach includes very low interference from the sample matrix, as the transducer is only sensitive to ferromagnetic substances which rarely are present in the sample [94, 95].

Very sensitive superconducting quantum interference device (SQUID) magnetometers have been tested to measure antigen-antibody interactions. In this system, antibodies are labelled with magnetic nanoparticles, and the antibody-antigen reaction is measured by detecting the magnetic field from the magnetic nanoparticles present in the complex. At present, 4×10^6 magnetic markers (diameter = 50 nm), corresponding to 520 pg of magnetic material, can be detected [96–98].

Modification, detection, isolation, and study of cells and cell organelles

The immunomagnetic separation of cells has become an important tool used especially in cell biology and medicine [59]. A substantial amount of experiments has been performed using magnetic nanoparticles bearing antibodies specific to the cell surface epitopes.

In general, two different modes of separation can be used. In the direct method, an appropriate affinity ligand coupled to magnetic nanoparticles is applied directly to the sample. During the incubation magnetic affinity particles are bound to the target cells, and thus stable magnetic complexes are formed. In the indirect method, a free affinity ligand (in most cases an appropriate antibody, often biotinylated) is first added to the cell suspension. If possible, the excess of unbound affinity ligand (antibody) is removed after incubation, and the labelled cells are then captured by magnetic nanoparticles bearing an affinity ligand against the primary label (*e.g.* secondary antibodies or streptavidin). In both methods the resulting magnetic complex is separated using an appropriate magnetic separator [59].

Both positive isolation (when target cellular subsets are magnetically labelled and subsequently separated) and negative isolation (when targets are purified by removing all other contaminating cells) can be performed. Cells of various types can be isolated using equipment and dextran-based magnetic particles from *e.g.* Miltenyi Biotec, Germany, or StemCell Technologies, USA.

Magnetic nanoparticles used to label the cells have no negative effect on the viability of the attached cells, and the isolated cells remain phenotypically unaltered. The extremely small size of the magnetic nanoparticles (*ca.* 50 nm) avoids mechanical stress for the cells and allows short incubation and fast processing times. The particles form a stable colloidal suspension and do not sediment or aggregate in magnetic fields. Their size and composition (iron oxide and polysaccharide) make the particles biodegradable, and typically they do not activate cells or influence cell functions and viability. Cells retain their physiological function. Nanoparticles detachment is not required, so positively selected cells (*i.e.* magnetically labelled ones) can be used immediately after separation for analysis and subsequent experiments. Magnetically labelled cells can be simultaneously stained by fluorochrome conjugated antibodies, facilitating quality control and analysis of the separation. The bound nanoparticles do not affect the light scattering of labelled cells. The purity of the sorted fractions can be determined directly after magnetic separation by flow cytometry. The separation procedure is also compatible with fluorescence microscopy, PCR, or FISH.

Very pure cell populations with excellent recovery and viability can be isolated; typical purities reach 95–99.9%, and >90% recovery can be achieved depending

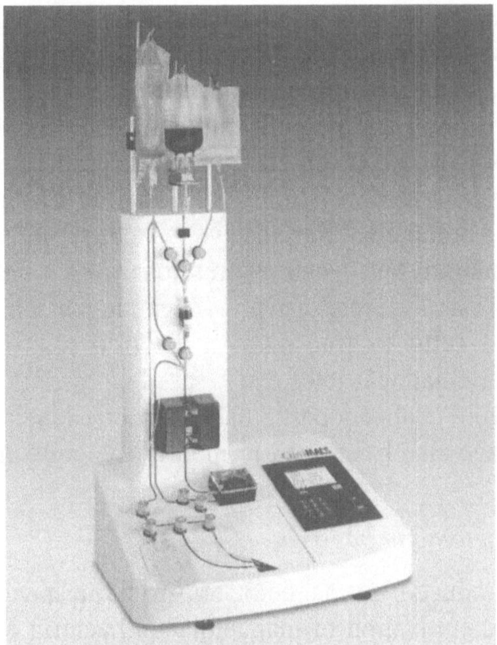

Fig. 5. The automated system for clinical isolation of human cell subsets CliniMACS. Reproduced with permission from materials provided by Miltenyi Biotec, Germany

on the cell frequency and the level of marker expression. Larger amounts of cells can be isolated using automated systems developed by Miltenyi Biotec, Germany; CliniMACS (Fig. 5) allows cells to be separated in a closed, sterile system, whereas AutoMACS enables high throughput sample separation, *e.g.* for further flow cytometric analysis or sorting.

A great variety of cell types has been isolated up to now. Especially important is the process of detection and removal of circulating tumour cells using an immunomagnetic procedure [99]. Another important process is the selective separation of CD34+ cells (stem cells) which opens new possibilities for stem cell transplantation and genetic manipulation of the hematopoietic system [100]. Nanoparticles obtained from magnetotactic bacteria with immobilized antibodies have been used for the detection and removal of *Escherichia coli* [67].

Annexin V is a Ca^{2+}-dependent phospholipid binding protein with high affinity for phosphatidylserine (*PS*) which is redistributed from the inner to the outer plasma membrane leaflet in early apoptosis. If immobilized to magnetic nanoparticles it can be used for the purification of apoptotic and non-apoptotic cells [32].

Not only whole cells, but also various cell organelles can be selectively separated using magnetic biocompatible nanoparticles. The lysosome fraction was isolated from the amoeba *Dictyostelium discoideum* after feeding with dextran-based nanoparticles and subsequent homogenization [101, 102]. A similar procedure was used to isolate lysosomes from human epidermal keratinocytes [103]. Plasma membranes from Chinese hamster ovary cells have been isolated after binding of wheat germ agglutinin immobilized on magnetic nanoparticles to the cell surface, followed by cell disintegration and magnetic separation [104].

A complex of magnetic cationic liposomes and a plasmid was used for transfection of selected animal cells. After incubation, the cells were subjected to magnetic separation, and transformed cells were selected. No other marker is required, and the separation can be performed just after the transformation [105].

Bacterial magnetite particles of 50 to 100 nm diameter were used as *DNA* carriers for the ballistic transformation of the marine cyanobacterium *Synechococcus*. Particles were bombarded into the cyanobacterial cells using a particle gun. Successful transformation and gene expression were confirmed by Southern hybridization and CAT assays, respectively. Magnetic particles were also observed in the cyanobacterial cells by transmission electron microscopy. These results suggested that magnetic nanoparticles can be used as carriers for introducing *DNA* into bacterial cells [106]. Subnanoparticulate magnetic labels such as erbium ions or magnetoferritin have also been used to modify the cell surface [59].

Applications of magnetotactic bacteria

Magnetic particles produced by magnetotactic bacteria have found various practical applications, and cultivation of magnetotactic bacteria can thus be an important process for the production of fine-grade, homogeneous, and biocompatible magnetite under mild conditions at normal temperature and pressure. Only a limited number of magnetotactic bacteria have been isolated in pure culture so far. For larger-scale cultivation, aerotolerant strains are preferred. *Magnetospirillum gryphiswaldense* and *Magnetospirillum* AMB-1 are tolerant to atmospheric air if grown from large inocula. The maximum cell yields reported so far are 0.34 g/dm^3 (dry weight; this corresponded to 4.5 mg of bacterial magnetic particles) for *Magnetospirillum* AMB-1 grown using a fed-batch culture system in a 4 dm^3 fermentor [107] and 0.33 g/dm^3 (dry weight) for *M. gryphiswaldense* grown in a 100 dm^3 fermenter [108]. Magnetic particles are usually released from the cells after disruption by ultrasonication or French press. Subsequent techniques for the isolation and purification of magnetosome particles from *Magnetospirillum* species are based on magnetic separation [109, 110] or a combination of a sucrose-gradient centrifugation and a magnetic separation technique [111]. These procedures leave the surrounding membrane intact, and magnetosome preparations are apparently free of contaminating material. Owing to the presence of the enveloping membrane, isolated magnetosome particles form stable, well-dispersed suspensions [7].

Both the cells of magnetotactic bacteria and the isolated magnetic particles have found interesting applications. The cells were used for the nondestructive domain analysis of soft magnetic materials [112] or to locate magnetic poles on meteoritic magnetic grains [113]. Experiments with possible applications of magnetotactic bacteria for radionuclide recovery have also been performed [114, 115]. In the near future, cultivation of genetically engineered magnetotactic bacteria producing magnetic nanoparticles with attached specific proteins can be expected.

Drug and radionuclide targeting

Drug and radionuclide targeting, *i.e.* predominant active compound accumulation in the body target zone independent from the method and route of drug administration, may resolve problems currently associated with systemic drug administration. One of the possible schemes for drug targeting includes magnetic targeting. For this purpose, the drug or radionuclide can be immobilized in biocompatible magnetic nano- or microspheres or in magnetoliposomes. Typically, the intended drug and an appropriate ferrofluid are formulated into a pharmaceutically stable formulation. This is then usually injected through the artery that supplies the target organ or tumor in the presence of an external magnetic field. Prolonged retention of the magnetic drug carrier at the target site alleviates or delays the RES clearance and facilitates extravascular uptake. This process is based on competition between forces exerted on the particles by the macro- and microcirculation, the characteristics of the magnetic particles (size, configuration), and the applied magnet. To effectively retain the magnetic drug carrier, the magnetic forces must be high enough to counteract linear flow rates within the organ or tumour tissue (between 10 and 0.05 cm/s depending on vessel size and branching patterns) [116–118].

Current technologies of magnetic drug targeting allow the localization of up to 70% of the administrated dose in the target tissue, with minimal interaction and toxicity to normal cells. An up to eight-fold increase in drug concentration in the target tissue after administration of only a third of the drug dose has been observed [116].

Special types of ferrofluids stabilized with anhydroglucose polymers have been developed which enabled chemoadsorptive binding of various drugs, cytokines, *DNA* fragments, and other molecules. Animal studies have demonstrated good tolerance of the magnetically targeted ferrofluid: low concentrations of the magnetic fluid delivered high drug doses and allowed for effective tumor therapy. Clinical studies have been performed on patients with various types of tumors using a magnetic fluid with bound epirubicin. The amount of ferrofluid applied to the patients was 0.5% of the estimated blood volume. The ferrofluid was well tolerated, and in some of the patients the therapeutic procedure was partially effective [116, 119].

Magnetic fluids can be used to prepare various types of magnetic biocompatible and biodegradable (bio)polymer particles and magnetoliposomes that can be used to encapsulate various drugs and radionuclides [120, 121]. Thermosensitive magnetoliposomes can release the entrapped drugs after selective heating caused by an electromagnetic field [122, 123].

In animal experiments, the local prevention of thrombosis in arteries of dogs and rabbits has been achieved by the intravenous application of the autologous red blood cells loaded with ferromagnetic colloid suspension and aspirin if a strong SmCo magnet was secured externally to the artery where the thrombus was initiated [124].

Magnetic fluid hyperthermia

During cancer therapy many procedures have been used. Hyperthermia is a promising approach to cancer therapy based on the heating of the target tissue to temperatures between 42 and 46°C, thus generally reducing the viability of cancer cells and increasing their sensitivity to chemotherapy and radiation.

Unlike chemotherapy and radiotherapy, hyperthermia itself has fewer side effects. Magnetic fluid hyperthermia is based on the fact that subdomain magnetic particles produce heat through various kinds of energy losses during application of an external AC magnetic field. If magnetic particles can be accumulated only in the tumor tissue, cancer specific heating is available.

In 1979, *Gordon et al.* [125] have used for the first time a magnetic fluid based on dextran magnetite nanoparticles to treat mammary tumour bearing rats. Since then, various types of biocompatible magnetic fluids [126, 127], cationic magnetoliposomes [128], or affinity magnetoliposomes [129] have been used for hyperthermia treatment, and several review papers are available on this topic [127, 130].

An interesting possibility of cancer treatment is the combination of hyperthermia treatment followed by chemotherapy or gene therapy. In this case, magnetic nano- or microspheres or magnetoliposomes containing a drug are first used to cause hyperthermia using the standard procedure; subsequently, the released drug acts on the injured cancer cells. This combined treatment might be very efficient [131]. Alternatively, magnetoliposomes have been used to cause hyperthermia which also resulted in the expression of interferon-β from the gene under the control of a heat inducible promoter, inserted to the tumors cells using standard liposomes [132].

At present, systems for magnetic fluid hyperthermia therapy are under development, and the phase I of the clinical testing is under preparation [127].

Contrast-increasing materials during magnetic resonance imaging

Currently, magnetic resonance imaging (MRI) belongs to standard medical examination methods. MRI is essentially proton NMR performed on tissue. In MRI, image contrast is a result of the different signal intensity each tissue produces in response to a particular sequence of the applied radiofrequency pulses. This response depends on proton density and magnetic relaxation times so as MRI contrast depends on the chemical composition (especially on the concentration of water or lipid molecules) and molecular structure of the tissue and is usually manipulated by adjusting the instrumental parameters [133]. In early 1980s it was recognized that target-specific superparamagnetic particles can serve as a dramatic source of exogenous contrast and have rapidly become an important and indispensable tool for the non-invasive study of biological processes with MRI. Superparamagnetic magnetite-dextran nanoparticles change the rate at which protons decay from their excited state to the ground state. As a result, regions containing the superparamagnetic contrast agent appear darker in an MRI than regions without the agent. For instance, when superparamagnetic nanoparticles are delivered to the liver, healthy liver cells can uptake the particles whereas diseased cells cannot. Consequently, the healthy regions are darkened, and the diseased regions remain bright [134]. A novel use of these nanoparticles deals with tracking cells *in vivo*. Rat T-cells were labelled with superparamagnetic Fe_3O_4 nanoparticles. Inflamed rats' testicles attracted the magnetite-attached T-cells and caused a decrease in the MRI signal from the testicles [135]. Alternatively, antibody-conjugated magnetite nanoparticles can serve as a target-directed magnetic resonance contrast agent [136]. Polyethylene glycol-modified magnetoliposomes with a diameter of 40 nm and containing 1–6 superparamagnetic iron oxide crystals per vesicle have been found to have excellent properties as a bone marrow-seeking MR contrast agent [137].

Miscellaneous and potential medical applications

Recently, the direct antitumour effect of biocompatible cobalt-ferrite-based magnetic fluid on bitch mammary tumor cells has been studied. It was observed that tumor cells intensively endocytosed magnetic nanoparticles and became overloaded by them. This situation led to a massive necrosis of tumour cells [138].

Two methods have been proposed for the treatment of the acquired immunodeficiency syndrome (AIDS) by introducing magnetoliposomes with coupled human immunodeficiency virus (HIV) receptor proteins into the blood stream. In the first procedure, HIV bound to magnetoliposomes should be separated from the patient's body using the arterio-venous shunt connected with a high gradient magnetic separation device where the complex should be removed [139]. The second procedure is based on the inductive heating of the HIV-magnetoliposome complexes in an AC magnetic field which should inactivate the AIDS virus [140].

Silicone magnetic fluid has been suggested for use in eye surgery to help in the course of retinal detachment repair. The procedure should employ a magnetized encircling scleral buckle holding in place a magnetic fluid providing 360° encircling internal tamponade [141, 142].

A possible procedure leading to the damage of the target cells has been described. Magnetic nanoparticles complexed with an appropriate antibody will bind to the cell membrane of the target cell. After application of an external rotating magnetic field of rapidly changing polarity, the ferrofluid particles will be drawn into a circular path, and an axial spin will be induced as each particle aligns itself with the magnetic force lines. A portion of these magnetic fluid particles will be drawn into the target cell membrane and into the cytoplasm causing brief perforations of the cell membrane of the target cells. If enough mechanical damage is done to the plasma membrane or to the intracellular structures, cell lysis may result, but in any case the brief disruptions of the target cell membrane can be used to selectively introduce membrane impermeant cytotoxic or antiretroviral substances into the target cell while relatively sparing normal cells [143].

A brain tumor position sensing method using a magnetoimpedance micromagnetic sensor in combination with magnetic fluids has been proposed. Sensing of magnetic fields generated from magnetic nanoparticles accumulated in the tumor tissue could be an effective way to accurately detect the tumor position during the surgical operation [144].

An artificial sphincter muscle using magnetic fluid has been developed in Japan for use in an artificial anus. The device has been used successfully in dogs, and application in humans is foreseen [26]. Ferrofluid has also been used as a seal for an axial flow pump during the development of an implantable artificial heart [145].

Applications in other biosciences

The influence of biocompatible magnetic fluids on the physiological functions of various plants have been tested. The presence of magnetic nanoparticles in growth media caused different kinds of modifications in plant growth, organogenesis, life cycle, and cell structure. For example, root and leaf induction was accelerated in 2–6 days in the presence of ferrofluid. Positive effect of ferrofluids have also been observed in *in vitro* plant regenerates which were grown in hypogravity conditions [146, 147].

Magnetic field sensitive polymer gels containing magnetic nanoparticles exhibit a quick controllable change of shape caused by the change of magnetic field. This property can be used to mimic muscular contractions. The peculiar magneto-elastic properties of these gels may be used to create a wide range of motion and to control a smooth and gentle shape change and movement similar to what is observed in muscles [148].

Future Trends

As can be seen, magnetic nanoparticles represent an extremely interesting group of inorganic nanomaterials, having close connections to living systems and their components. Their importance has even exceeded the planet Earth, and magnetic nanoparticles might be among the first proofs of the presence of extraterrestrial life.

Further studies of magnetic nanoparticles biomineralization processes will be interesting not only from the point of view of basic research, but also with respect to large-scale synthesis of magnetic biocompatible nanoparticles. Biotechnology production of either unmodified or genetically engineered bacterial magnetosomes may be of great interest and become one of the standard processes for magnetic nanoparticle preparation.

Separation processes employing magnetic modification of originally diamagnetic components followed by magnetic separation will be used more frequently. These techniques will become standard procedures in biology and clinical laboratories.

Magnetic assays, especially those employing magnetic nanoparticles as specific labels, will certainly find more applications in the near future. The progress in this area will be supported by the further development of computer hardware technology which will enable the detection of minute amounts of magnetic labels.

Probably the most important applications of magnetic nanoparticles in the area of biosciences can be expected in medicine and related disciplines. Magnetic drugs, antibiotics, radionuclides, genes, *etc.* targeting, magnetic fluid hyperthermia, detection of cancer cells, isolation of stem cells, possible influence on biological functions by specific types of ferrofluids, improvement of diagnostic procedures (such as MRI), development of clinical biochemistry assays based on the application of magnetic nanoparticles – these all and most probably several other procedures will be further developed to employ the unique properties of magnetic nanoparticles. Not only the individual procedures, but also their combinations (such as magnetic drug targeting combined with hyperthermia) can lead to very interesting results.

The combination of nanotechnologies and biosciences will be one of the leading areas of research and development in the 21st century; magnetic nanoparticles will certainly play an extremely important role.

Acknowledgements

This work was supported by the Ministry of Education of the Czech Republic (grant project No. OC 523.80) and the NATO Science Programme (Collaborative Linkage Grant No. LST.CLG.977500).

References

[1] Huczko A (2000) Appl Phys A – Mater Sci Process **70**: 365
[2] Raj K, Moskowitz B, Casciari R (1995) J Magn Magn Mater **149**: 174
[3] Rosensweig RE (1982) Sci Amer **247**: 136
[4] Lowenstam HA (1962) Bull Geol Soc Am **73**: 435
[5] Kirschvink JL, Hagadorn JW (2000) In: Bauerlein E (ed) The Biomineralisation of Nano- and Micro-Structures. Wiley-VCH, Weinheim, p 139
[6] Blakemore R (1975) Science **190**: 377
[7] Schüler D, Frankel RB (1999) Appl Microbiol Biotechnol **52**: 464
[8] Matsunaga T, Sakaguchi T (2000) J Biosci Bioeng **90**: 1
[9] Schüler D (2000) In: Bartlett DH (ed) Molecular Marine Microbiology. Horizon Scientific Press, Wymondham, p 157
[10] Zhang CL, Vali H, Romanek CS, Phelps TJ, Liu SV (1998) Am Mineral **83**: 1409
[11] Torres de Araujo FF, Pires MA, Frankel RB, Bicudo CEM (1986) Biophys J **50**: 375
[12] Peck JA, King JW (1996) Earth Planet Sc Lett **140**: 159
[13] Wajnberg E, Acosta-Avalos D, El-Jaick LJ, Abracado L, Coelho JLA, Bakuzis AF, Morais PC, Esquivel DMS (2000) Biophys J **78**: 1018
[14] El-Jaick LJ, Acosta-Avalos D, De Souza DM, Wajnberg E, Linhares MP (2001) Eur Biophys J Biophys Lett **29**: 579
[15] Winklhofer M, Holtkamp-Rotzler E, Hanzlik M, Fleissner G, Petersen N (2001) Eur J Mineral **13**: 659
[16] Ogura M, Kato M, Arai N, Sasada T, Sakaki Y (1992) Can J Zool **70**: 874
[17] Lohmann KJ, Johnsen S (2000) Trends Neurosci **23**: 153
[18] Kirschvink JL, Walker MM, Diebel CE (2001) Curr Opin Neurobiol **11**: 462
[19] Acosta-Avalos D, Wajnberg E, Oliveira PS, Leal I, Farina M, Esquivel DMS (1999) J Exp Biol **202**: 2687
[20] Walker MM, Diebel CE, Haugh CV, Pankhurst PM, Montgomery JC, Green CR (1997) Nature **390**: 371
[21] Diebel CE, Proksch R, Green CR, Neilson P, Walker MM (2000) Nature **406**: 299
[22] Dobson J (2001) FEBS Lett **496**: 1
[23] Thomas-Keprta KL, Bazylinski DA, Kirschvink JL, Clemett SJ, McKay DS, Wentworth SJ, Vali H, Gibson EK, Romanek CS (2000) Geochim Cosmochim Acta **64**: 4049
[24] Friedmann EI, Wierzchos J, Ascaso C, Winklhofer M (2001) Proc Natl Acad Sci USA **98**: 2176
[25] Weiss BP, Kirschvink JL, Baudenbacher FJ, Vali H, Peters NT, Macdonald, FA, Wikswo JP (2000) Science **290**: 791
[26] Berkovski B, Bashtovoy V (eds) (1996) Magnetic Fluids and Applications Handbook. Begell House, New York
[27] Massart R (1981) IEEE Trans Magn **17**: 1247
[28] Berger P, Adelman NB, Beckman KJ, Campbell DJ, Ellis AB, Lisensky GC (1999) J Chem Educ **76**: 943

[29] Shen LF, Laibinis PE, Hatton TA (1999) J Magn Magn Mater **194**: 37
[30] Shen LF, Laibinis PE, Hatton TA (1999) Langmuir **15**: 447
[31] Sousa MH, Rubim JC, Sobrinho PG, Tourinho FA (2001) J Magn Magn Mater **225**: 67
[32] Halbreich A, Roger J, Pons JN, Geldwerth D, DaSilva MF, Roudier M, Bacri JC (1998) Biochimie **80**: 379
[33] Domingo JC, Mercadal M, Petriz J, De Madariaga MA (2001) J Microencapsul **18**: 41
[34] Tiefenauer LX, Kuhne G, Andres RY (1993) Bioconjugate Chem **4**: 347
[35] Molday RS, Mackenzie D (1982) J Immunol Methods **52**: 353
[36] Pardoe H, Chua-anusorn W, Pierre TGS, Dobson J (2001) J Magn Magn Mater **225**: 41
[37] O'Connor CJ, Seip CT, Carpenter EE, Li SC, John VT (1999) Nanostruct Mater **12**: 65
[38] O'Connor CJ, Kolesnichenko V, Carpenter E, Sangregorio C, Zhou WL, Kumbhar A, Sims J, Agnoli F (2001) Synthetic Met **122**: 547
[39] Seip CT, O'Connor CJ (1999) Nanostruct Mater **12**: 183
[40] De Cuyper M, Joniau M (1988) Eur Biophys J **15**: 311
[41] De Cuyper M, Joniau M (1993) J Magn Magn Mater **122**: 340
[42] Rocha FM, de Pinho SC, Zollner RL, Santana MHA (2001) J Magn Magn Mater **225**: 101
[43] Sangregorio C, Wiemann JK, O'Connor CJ, Rosenzweig Z (1999) J Appl Phys **85**: 5699
[44] Menager C, Cabuil V (1994) Colloid Polym Sci **272**: 1295
[45] Mosbach K, Andersson L (1977) Nature **270**: 259
[46] Prestvik WS, Berge A, Mork PC, Stenstad PM, Ugelstad J (1997) In: Häfeli U, Schütt W, Teller J, Zborowski M (eds) Scientific and Clinical Applications of Magnetic Carriers. Plenum, New York London, p 11
[47] Yanase N, Noguchi H, Asakura H, Suzuta T (1993) J Appl Polym Sci **50**: 765
[48] Ding XB, Sun ZH, Wan GX, Jiang YY (1998) React Funct Polym **38**: 11
[49] Abe M (2000) Electrochim Acta **45**: 3337
[50] Margel S, Gura S, Bamnolker H, Nitzan B, Tennenbaum T, BarToov B, Hinz M, Seliger H (1997) In: Häfeli U, Schütt W, Teller J, Zborowski M (eds) Scientific and Clinical Applications of Magnetic Carriers. Plenum, New York London, p 37
[51] Sauzedde F, Elaissari A, Pichot C (1999) Colloid Polym Sci **277**: 846
[52] Sauzedde F, Elaissari A, Pichot C (1999) Colloid Polym Sci **277**: 1041
[53] Ziolo RF, Giannelis EP, Weinstein BA, Ohoro MP, Ganguly BN, Mehrotra V, Russell MW, Huffman DR (1992) Science **257**: 219
[54] Llanes F, Ryan DH, Marchessault RH (2000) Int J Biol Macromol **27**: 35
[55] Marchessault RH, Ricard S, Rioux P (1992) Carbohyd Res **224**: 133
[56] Hirano S, Yogo T, Sakamoto W, Yamada S, Nakamura T, Yamamoto T, Ukai H (2001) J Eur Ceram Soc **21**: 1479
[57] Matsunaga T, Takeyama H (1998) Supramol Sci **5**: 391
[58] Meldrum FC, Heywood BR, Mann S (1992) Science **257**: 522
[59] Šafařík I, Šafaříková M (1999) J Chromatogr B **722**: 33
[60] Šafařík I, Šafaříková M, Forsythe SJ (1995) J Appl Bacteriol **78**: 575
[61] Šafaříková M, Šafařík I (2001) Magn Electr Sep **10**: 223
[62] Šafařík I, Šafaříková M (1997) In: Häfeli U, Schütt W, Teller J, Zborowski M (eds) Scientific and Clinical Applications of Magnetic Carriers. Plenum, New York London, p 323
[63] Cell Separation and Protein Purification, Information booklet, Dynal, Oslo, Norway, 1996
[64] Biomagnetic Techniques in Molecular Biology, Information booklet, Dynal, Oslo, Norway, 1998
[65] Matsunaga T, Kamiya S (1987) Appl Microbiol Biotechnol **26**: 328
[66] Matsunaga T, Kawasaki M, Yu X, Tsujimura N, Nakamura N (1996) Anal Chem **68**: 3551
[67] Nakamura N, Burgess JG, Yagiuda K, Kudo S, Sakaguchi T, Matsunaga T (1993) Anal Chem **65**: 2036
[68] Nakamura N, Matsunaga T (1993) Anal Chim Acta **281**: 585

[69] Matsunaga T, Nakayama H, Okochi M, Takeyama H (2001) Biotechnol Bioeng **73**: 400
[70] Sode K, Kudo S, Sakaguchi T, Nakamura N, Matsunaga T (1993) Biotechnol Tech **7**: 688
[71] Sestier C, DaSilva MF, Sabolovic D, Roger J, Pons JN (1998) Electrophoresis **19**: 1220
[72] Halbreich A, Roger J, Pons JN, da Silva MF, Hasmonay E, Roudier M, Boynard M, Sestier C, Amri A, Geldwerth D, Fertil B, Bacri JC, Sabolovic D (1997) In: Häfeli U, Schütt W, Teller J, Zborowski M (eds) Scientific and Clinical Applications of Magnetic Carriers. Plenum, New York London, p 399
[73] Shinkai M, Honda H, Kobayashi T (1991) Biocatalysis **5**: 61
[74] Shi YJ, Shen RN, Lu L, Broxmeyer HE (1994) Blood Cells **20**: 517
[75] Tamaura Y, Takahashi K, Kodera Y, Saito Y, Inada Y (1986) Biotechnol Lett **8**: 877
[76] Yoshimoto T, Mihama T, Takahashi K, Saito Y, Tamaura Y, Inada Y (1987) Biochem Biophys Res Commun **145**: 908
[77] Yoshimoto T, Ohwada K, Takahashi K, Matsushima A, Saito Y, Inada Y (1988) Biochem Biophys Res Commun **152**: 739
[78] Suzuki M, Shinkai M, Kamihira M, Kobayashi T (1995) Biotechnol Appl Biochem **21**: 335
[79] De Cuyper M, De Meulenaer B, Van der Meeren P, Vanderdeelen J (1995) Biocatal Biotransform **13**: 77
[80] Shinkai M, Suzuki M, Iijima S, Kobayashi T (1995) Biotechnol Appl Biochem **21**: 125
[81] Babincová M (1995) Pharmazie **50**: 702
[82] Matsunaga T, Kamiya S, Tsujimura N (1997) In: Häfeli U, Schütt W, Teller J, Zborowski M (eds) Scientific and Clinical Applications of Magnetic Carriers. Plenum, New York London, p 287
[83] Tanaka T, Matsunaga T (2000) Anal Chem **72**: 3518
[84] Matsunaga T, Sato R, Kamiya S, Tanaka T, Takeyama H (1999) J Magn Magn Mater **194**: 126
[85] Matsunaga T, Togo H, Kikuchi T, Tanaka T (2000) Biotechnol Bioeng **70**: 704
[86] Sakamoto A, Abe M, Masaki T (1999) FEBS Lett **447**: 124
[87] Shinkai M, Kamihira M, Honda H, Kobayashi T (1992) Kagaku Kogaku Ronbunshu **18**: 256
[88] Griffin T, Mosbach K, Mosbach R (1981) Appl Biochem Biotechnol **6**: 283
[89] Wikstrom P, Flygare S, Grondalen A, Larsson PO (1987) Anal Biochem **167**: 331
[90] Suzuki M, Kamihira M, Shiraishi T, Takeuchi H, Kobayashi T (1995) J Ferment Bioeng **80**: 78
[91] Shinkai M, Wang J, Kamihira M, Iwata M, Honda H, Kobayashi T (1992) J Ferment Bioeng **73**: 166
[92] Sato R, Takeyama H, Tanaka T, Matsunaga T (2001) Appl Biochem Biotechnol **91–93**: 109
[93] Takeyama H, Tsuzuki H, Chow S, Nakayama H, Matsunaga T (2000) Marine Biotechnol **2**: 309
[94] Kriz CB, Radevik K, Kriz D (1996) Anal Chem **68**: 1966
[95] Kriz K, Gehrke J, Kriz D (1998) Biosens Bioelectron **13**: 817
[96] Enpuku K, Minotani T (2001) IEICE Trans Electron **E84C**: 43
[97] Enpuku K, Minotani T, Gima T, Kuroki Y, Itoh Y, Yamashita M, Katakura Y, Kuhara S (1999) Jpn J Appl Phys **38**: L1102
[98] Enpuku K, Hotta M, Nakahodo A (2001) Physica **C 357–360**: 1462
[99] Bilkenroth U, Taubert H, Riemann D, Rebmann U, Heynemann H, Meye A (2001) Int J Cancer **92**: 577
[100] Kato K, Radbruch A (1993) Cytometry **14**: 384
[101] Rodriguez-Paris J, Nolta KV, Steck TL (1993) J Biol Chem **268**: 9110
[102] Temesvari L, Rodriguez-Paris J, Bush J, Steck TL, Cardelli J (1994) J Biol Chem **269**: 25719
[103] Glombitza GJ, Becker E, Kaiser HW, Sandhoff K (1997) J Biol Chem **272**: 5199
[104] Warnock DE, Roberts C, Lutz MS, Blackburn WA, Young WWJ, Baenziger JU (1993) J Biol Chem **268**: 10145
[105] Nagatani N, Shinkai M, Honda H, Kobayashi T (1998) Biotechnol Tech **12**: 525

[106] Takeyama H, Yamazawa A, Nakamura C, Matsunaga T (1995) Biotechnol Tech **9**: 355
[107] Matsunaga T, Tsujimura N, Kamiya S (1996) Biotechnol Tech **10**: 495
[108] Schüler D, Baeuerlein E (1997) J Phys IV **7**: 647
[109] Gorby YA, Beveridge TJ, Blakemore RP (1988) J Bacteriol **170**: 834
[110] Okuda Y, Denda K, Fukumori Y (1996) Gene **171**: 99
[111] Schüler D, Baeuerlein E (1997) In: Trautwein A (ed) Bioinorganic Chemistry: Transition Metals in Biology and Coordination Chemistry. Wiley-VCH, Weinheim, p 24
[112] Harasko G, Pfutzner H, Futschik K (1995) IEEE Trans Magn **31**: 938
[113] Funaki M, Sakai H, Matsunaga T, Hirose S (1992) Phys Earth Planet Inter **70**: 253
[114] Bahaj AS, James PAB, Moeschler FD (1998) J Magn Magn Mater **177**: 1453
[115] Bahaj AS, Croudace IW, James PAB, Moeschler FD, Warwick PE (1998) J Magn Magn Mater **184**: 241
[116] Lubbe AS, Bergemann C (1997) In: Häfeli U, Schütt W, Teller J, Zborowski M (eds) Scientific and Clinical Applications of Magnetic Carriers. Plenum, New York London, p 457
[117] Widder KJ, Senyei AE (1983) Pharmacol Therapeut **20**: 377
[118] Torchilin VP (2000) Eur J Pharm Sci **11**: S81
[119] Lubbe AS, Bergemann C, Riess H, Schriever F, Reichardt P, Possinger K, Matthias M, Dorken B, Herrmann F, Gurtler R, Hohenberger P, Haas N, Sohr R, Sander B, Lemke AJ, Ohlendorf D, Huhnt W, Huhn D (1996) Cancer Res **56**: 4686
[120] Pulfer SK, Gallo JM (1997) In: Häfeli U, Schütt W, Teller J, Zborowski M (eds) Scientific and Clinical Applications of Magnetic Carriers. Plenum, New York London, p 445
[121] Kuznetsov AA, Filippov VI, Alyautdin RN, Torshina NL, Kuznetsov OA (2001) J Magn Magn Mater **225**: 95
[122] Viroonchatapan E, Sato H, Ueno M, Adachi I, Tazawa K, Horikoshi I (1997) J Control Release **46**: 263
[123] Babincová M, Babinec P (1997) Cell Mol Biol Lett **2**: 3
[124] Orekhova NM, Akchurin RS, Belyaev AA, Smirnov MD, Ragimov SE, Orekhov AN (1990) Thromb Res **57**: 611
[125] Gordon RT, Hines JR, Gordon D (1979) Med Hypotheses **5**: 83
[126] Brusentsov NA, Gogosov VV, Brusentsova TN, Sergeev AV, Jurchenko NY, Kuznetsov AA, Kuznetsov OA, Shumakov LI (2001) J Magn Magn Mater **225**: 113
[127] Jordan A, Scholz R, Maier-Hauff K, Johannsen M, Wust P, Nadobny J, Schirra H, Schmidt H, Deger S, Loening S, Lanksch W, Felix R (2001) J Magn Magn Mater **225**: 118
[128] Yanase M, Shinkai M, Honda H, Wakabayashi T, Yoshida J, Kobayashi T (1998) Jpn J Cancer Res **89**: 463
[129] Le B, Shinkai M, Kitade T, Honda H, Yoshida J, Wakabayashi T, Kobayashi T (2001) J Chem Eng Jpn **34**: 66
[130] Jordan A, Scholz R, Wust P, Fahling H, Felix R (1999) J Magn Magn Mater **201**: 413
[131] Shinkai M, Suzuki M, Yokoi N, Yanase M, Shimizu W, Honda H, Kobayashi T (1998) Jpn J Hyperthermic Oncol **14**: 15
[132] Bouhon IA, Shinkai M, Honda H, Mizuno M, Wakabayashi T, Yoshida J, Kobayashi T (1999) Cancer Lett **139**: 153
[133] Bulte JWM, Brooks RA (1997) In: Häfeli U, Schütt W, Teller J, Zborowski M (eds) Scientific and Clinical Applications of Magnetic Carriers. Plenum, New York London, p 527
[134] Martin CR, Mitchell DT (1998) Anal Chem **70**: A322
[135] Yeh TC, Zhang WG, Ildstad ST, Ho C (1995) Magn Reson Med **33**: 200
[136] Suzuki M, Honda H, Kobayashi T, Wakabayashi T, Yoshida J, Takahashi M (1996) Brain Tumor Pathol **13**: 127
[137] Bulte JWM, de Cuyper M, Despres D, Frank JA (1999) J Magn Magn Mater **194**: 204
[138] Sincai M, Ganga D, Bica D, Vekas L (2001) J Magn Magn Mater **225**: 235
[139] Babincová M, Machová E (1998) Z Naturforsch **C 53**: 935

[140] Müller-Schulte D, Füssl F, Lueken H, De Cuyper M (1997) In: Häfeli U, Schütt W, Teller J, Zborowski M (eds) Scientific and Clinical Applications of Magnetic Carriers. Plenum, New York London, p 517
[141] Dailey JP, Phillips JP, Li C, Riffle JS (1999) J Magn Magn Mater **194**: 140
[142] Voltairas PA, Fotiadis DI, Massalas CV (2001) J Magn Magn Mater **225**: 248
[143] Peasley KW (1996) Med Hypotheses **46**: 5
[144] Uchiyama T, Mohri K, Shinkai M, Ohshima A, Honda H, Kobayashi T, Wakabayashi T, Yoshida J (1997) IEEE Trans Magn **33**: 4266
[145] Mitamura Y, Fujiyoshi M, Yoshida T, Yozu R, Okamoto E, Tanaka T, Kawada S (1996) Artif Organs **20**: 497
[146] Butnaru G, Terteac D, Potencz I (1999) J Magn Magn Mater **201**: 435
[147] Butnaru G (2000) In: Proc Conf "Propagation of Ornamental Plants", Sofia, p 35
[148] Zrinyi M (2000) Colloid Polym Sci **278**: 98
[149] Heywood BR, Bazylinski DA, Garrattreed A, Mann S, Frankel RB (1990) Naturwissenschaften **77**: 536
[150] Bazylinski DA, Schlezinger DR, Howes BH, Frankel RB, Epstein SS (2000) Chem Geol **169**: 319
[151] Maher BA (1998) Proc R Soc Lond Ser B – Biol Sci **265**: 733
[152] Moore A, Freake SM, Thomas IM (1990) Philos T Roy Soc B **329**: 11
[153] Walker MM, Quinn TP, Kirschvink JL, Groot C (1988) J Exp Biol **140**: 51
[154] Brassart J, Kirschvink JL, Phillips JB, Borland SC (1999) J Exp Biol **202**: 3155
[155] Beason RC, Dussourd N, Deutschlander ME (1995) J Exp Biol **198**: 141
[156] Beason RC, Semm P (1996) J Exp Biol **199**: 1241
[157] Hanzlik M, Heunemann C, Holtkamp-Rotzler E, Winklhofer M, Petersen N, Fleissner G (2000) Biometals **13**: 325
[158] Zoeger J, Dunn JR, Fuller M (1981) Science **213**: 892
[159] Kirschvink JL, Kobayashi-Kirschvink A, Woodford BJ (1992) Proc Natl Acad Sci USA **89**: 7683
[160] Schultheiss-Grassi PP, Dobson J (1999) Biometals **12**: 67
[161] Schultheiss-Grassi PP, Wessiken R, Dobson J (1999) Biochim Biophys Acta **1426**: 212
[162] Dobson J, Grassi P (1996) Brain Res Bull **39**: 255
[163] Grassi-Schultheiss PP, Heller F, Dobson J (1997) Biometals **10**: 351

Received October 4, 2001. Accepted November 19, 2001

Direct Synthesis of Nanocrystalline Hydroxyapatite by Hydrothermal Hydrolysis of Alkylphosphates

Jaroslav Cihlar*, and **Klara Castkova**

Department of Ceramics, Brno University of Technology, Cz-61669 Brno, Czech Republic

Summary. The influence of reaction conditions (temperature, type of catalyst, time) on the base-catalyzed reaction of mono-, di-, and trialkylphosphates (alkyl = methyl, ethyl, *i*-propyl, *n*-butyl) with Ca^{2+} ions and on the structure and composition of the reaction products was studied. The composition of the calcium phosphates depends mainly on the reaction temperature. At temperatures below 100°C, a nanocrystalline solid product transforming into dicalcium phosphate by heating (calcination) was found. Pure nanocrystalline hydroxyapatite was prepared by hydrothermal synthesis at 160°C from mono- and dialkylphosphates. The size of hydroxyapatite crystallites was about 1 nm, the particle size about 150 nm. Agglomerated particles of hydroxyapatite larger than 2 μm were prepared at 200°C. Hydrothermal reaction of trialkylphosphates with Ca^{2+} ions at 200°C produced $CaHPO_4$. The experimental results were used to propose a reaction mechanism for base-catalyzed hydrothermal reactions of alkylphosphates with Ca^{2+} ions.

Keywords. Ceramics; Total synthesis; Nanostructures; Reaction mechanism; Hydroxyapatite; Nanoparticles; Hydrothermal synthesis; Alkylphosphates.

Introduction

Natural calcium phosphates, in particular hydroxyapatites (*HA*), form the fundamental inorganic component of the skeletal systems of vertebrates. Synthetic hydroxyapatites have similar properties to those of natural hydroxyapatites and are therefore frequently used as biomaterial in orthopaedics and dentistry. Among the most frequent medical applications of hydroxyapatite are biologically active surface layers applied to biologically inert ceramic or metallic materials. The usual methods of applying such coatings, *i.e.* flame and plasma spraying, operate at temperatures that often decompose hydroxyapatite into calcium phosphates which are of low stability in the organism. Exploring new methods of applying hydroxyapatite coatings represents an important field in contemporary biomaterials research. Most of these methods, such as dip coating, spraying, and spin coating, employ liquid precursors of hydroxyapatite at normal temperature. Among the most interesting precursors of calcium phosphates or phosphate ceramics are

* Corresponding author. E-mail: cihlar@ro.vutbr.cz

alkylphosphates because of their easy preparation *via* the reaction of P_2O_5 with alcohols [1–3].

Alkylphosphates reacting with aqueous solutions of calcareous salts at normal temperature form gels or precipitates that transform into hydroxyapatite or a mixture of calcium phosphates ($Ca_2P_2O_7$, $Ca_3(PO_4)_2$) during heat treatment (sintering) [4–7]. Pure hydroxyapatites have been prepared by hydrothermal decomposition of inorganic phosphates (of ammonia or calcium) in the presence of Ca^{2+} ions at a temperature of 500°C. Usually, microcrystalline particles of 5–25 μm in size were obtained [8–10]. With the temperature reduced to just above 100°C, this hydrothermal decomposition yielded nanocrystalline hydroxyapatites whose particle size was between 15 and 300 nm [11–13]. The hydrothermal method was used successfully also in the synthesis of crystalline hydroxyapatite from trialkylphosphate and calcium acetate. Particles of hydroxyapatite prepared at 350°C were 10–30 nm in size [14].

From the above overview it can be seen that the hydrothermal reaction of alkylphosphates with Ca^{2+} ions at temperatures ranging from 100 to 200°C could yield pure nanocrystalline hydroxyapatite suitable for the preparation of bioactive pure hydroxyapatite coatings by the sol-gel method (dip coating, spraying). The present paper is concerned with the study of this problem.

Results and Discussion

Reaction of mono- and dialkylphosphates with calcium acetate at 60°C

Determination of the phase composition of the precipitates formed at 60°C by reaction of methyl-, ethyl-, *i*-propyl-, and *n*-butyl phosphate with calcium acetate under normal conditions by means of X-ray analysis was not possible. The spectra did not correspond to any of the expected products ($CaHPO_4$, $Ca_2P_2O_7$, $Ca_3(PO_4)_2$, *HA*). The calcination of precipitates yielded mostly $Ca_2P_2O_7$, a mixture of $Ca_2P_2O_7$ and $Ca_3(PO_4)_2$, or *HA*.

Figure 1 shows the IR spectra of precipitates formed by the reaction of mono- and diethylphosphate 1, 2, and 3 hours after the reaction. It is evident from Fig. 1 that after 1 hour of reaction the composition of the precipitate did not change. The band at 2970 cm^{-1} corresponding to the vibration of the CH_3 group in the P–OEt group shows that the precipitate contained part of the non-reacted P–OEt groups, which did not change with the reaction time.

Hydrothermal reaction of mono- and dialkylphosphates with calcium acetate at 160°C

Under conditions of hydrothermal synthesis at 160°C and 16 MPa the preparation of hydroxyapatite required 4 hours in all cases (Fig. 2). The size of crystallites calculated from RTG spectra was within a narrow interval from 1.0 to 1.1 nm. The size and shape of hydroxyapatite particles can be seen in Fig. 3. The particles are of almost globular shape. In the case of the product from methyl

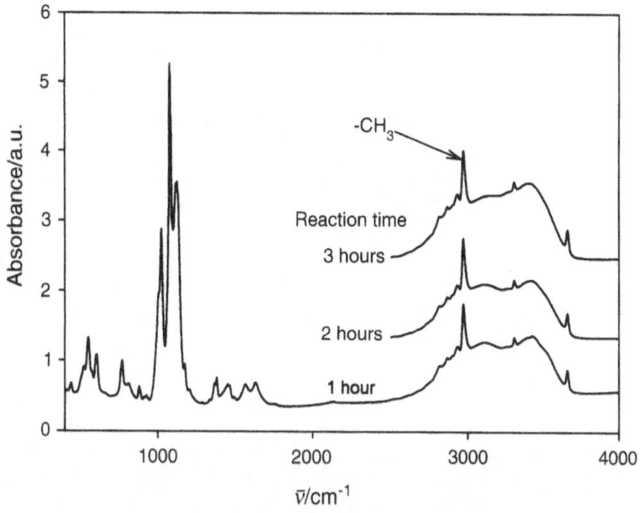

Fig. 1. FT-IR spectra of precipitates prepared by reaction of ethylphosphates with calcium acetate at 60°C for 1, 2 or 3 h

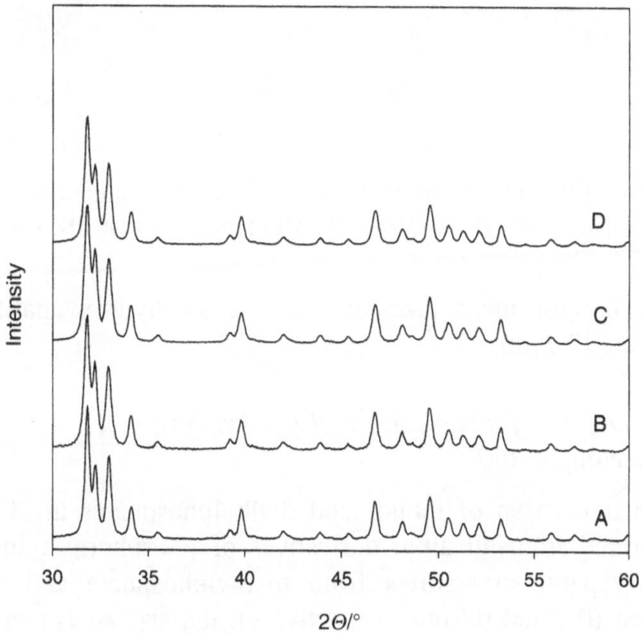

Fig. 2. XRD patterns of hydroxyapatite prepared by hydrothermal synthesis at 160°C from A) methylphosphate, B) ethylphosphate, C) *i*-propylphosphate, D) *n*-butylphosphate

phosphate the particle size was 180 nm, in the case of ethylphosphate 160 nm, in the case of *i*-propylphosphate 150 nm, and in the case of *n*-butylphosphate 160 nm (Table 1).

It follows from Fig. 3D that in particular in the case of hydroxyapatite prepared from *n*-butylphosphate the particles formed 'soft' agglomerates of more than 1 μm in size. The lowest mean size of agglomerates (measured by laser diffraction) was established in the case of hydroxyapatite prepared from methylphosphate

Fig. 3. SEM micrographs of hydroxyapatite particles prepared by hydrothermal synthesis at 160°C from A) methylphosphate, B) ethylphosphate, C) *i*-propylphosphate, D) *n*-butylphosphate

(0.59 μm), the highest mean size in the case of hydroxyapatite made from *i*-propylphosphate (2.2 μm).

Hydrothermal reaction of mono-, di-, and trialkylphosphates with calcium acetate at 200°C

Via hydrothermal reaction of mono- and dialkylphosphates at 200°C, hydroxyapatite was formed from all alkyl derivatives of phosphoric acid (Fig. 4). The size of hydroxyapatite crystallites from methylphosphate and ethylphosphate was the smallest (0.7 and 0.6 nm, respectively); the size of *HA* crystallites from *i*-propylphosphate and *n*-butylphosphate was 1.1 and 0.9 nm, respectively (Table 1, Fig. 4). The structure, size, and shape of *HA* particles are evident from Fig. 5. Hydroxyapatite particles formed from methyl- and ethylphosphate were of roughly globular shape, and their size was 230 and 210 nm, respectively. Hydroxyapatite particles formed from *i*-propyl- and *n*-butylphosphate had globular to elongated shape and 120 and 200 nm, respectively, in size. The particles of all hydroxyapatite products were agglomerated into 'hard' agglomerates whose sizes (established by laser diffraction) are given in Table 1. Agglomerates of methylphosphate were 2.74 μm in size, those of ethylphosphate 5.12 μm, those of *i*-propylphosphate 3.44 μm, and those of butylphosphate 4.95 μm. Breaking up the aggregates into separate particles could not be achieved by ultrasound but by colloidal milling for 6 hours.

Table 1. Results of analysis of Ca-phosphate products

Alkylphosphate temperature	Analysis			
	Phase composition	Size of prim. Crystallites (XRD)	Particle size (SEM)	Mean size of agglomerates (laser diffraction)
methylphosphate, 60°C	unknown	–	–	–
ethylphosphate, 60°C	unknown	–	–	–
i-propylphosphate, 60°C	unknown	–	–	–
n-butylphosphate, 60°C	unknown	–	–	–
methylphosphate, HT/160°C	*HA*	1.1 nm	180 nm	0.59 μm
ethylphosphate, HT/160°C	*HA*	1.0 nm	160 nm	1.31 μm
i-propylphosphate, HT/160°C	*HA*	1.0 nm	150 nm	2.22 μm
n-butylphosphate, HT/160°C	*HA*	1.0 nm	160 nm	1.70 μm
methylphosphate, HT/200°C	*HA*	0.7 nm	230 nm	2.74 μm
ethylphosphate, HT/200°C	*HA*	0.6 nm	210 nm	5.12 μm
i-propylphosphate, HT/200°C	*HA*	1.1 nm	120 nm	3.44 μm
n-butylphosphate, HT/200°C	*HA*	0.9 nm	200 nm	4.95 μm
trimethylphosphate HT/200°C	$CaHPO_4$	–	–	–
triethylphosphate HT/200°C	$CaHPO_4$	–	–	–

After hydrothermal reaction of trimethyl- and triethylphosphate with calcium acetate at 200°C the only crystalline phase found in the reaction product was $CaHPO_4$; no hydroxyapatite phase was detected.

Proposed reaction mechanism of hydrothermal synthesis of hydroxyapatite from alkylphosphates

The reaction of alkylphosphates with Ca^{2+} in aqueous medium proceeds in three basic steps. In the first step the hydrolysis of the P–O*R* bond takes place, in the second step the calcium salt of alkylphosphoric acid is formed, and in the third step the P–OH bond (formed by the hydrolysis of the P–O*R* bond) reacts with Ca^{2+} accompanied by the formation of Ca-phosphate [15]. Under normal temperatures the rate of the hydrolysis of P–O*R* bond is low [16, 17]. The reaction is reversible and, for example, in the product of the reaction of trialkylphosphate with Ca^{2+} one –O*R* group per one P was found [5]. These data from the literature are corroborated by our results obtained for the reaction of mono- and dialkylphosphates under

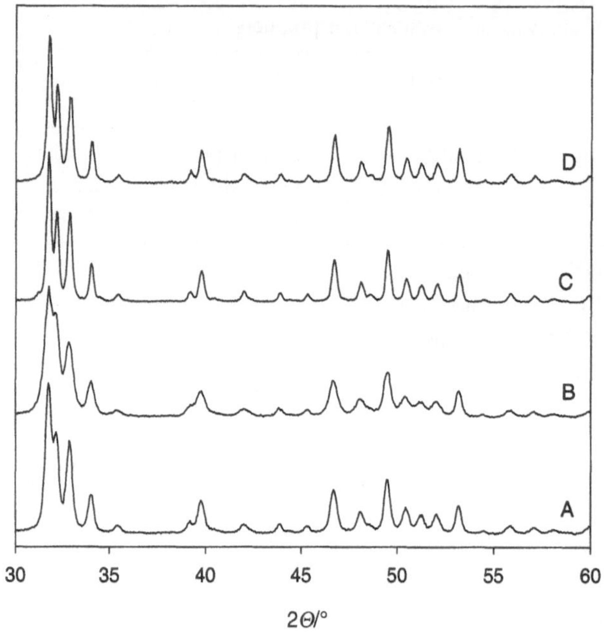

Fig. 4. XRD patterns of hydroxyapatite prepared by hydrothermal synthesis at 200°C from A) methylphosphate, B) ethylphosphate, C) *i*-propylphosphate, D) *n*-butylphosphate

normal conditions. The reaction product was not hydroxyapatite, but undefined crystalline compounds containing P–O*R* groups.

In the course of the hydrothermal reaction the rate of the hydrolysis of P–O*R* bonds increased, the stability of the P–O*R* bond was reduced, and the equilibrium was shifted towards the hydrolysis product. The base-catalyzed hydrolysis of the P–O*R* bond probably proceeds according to Eq. (1):

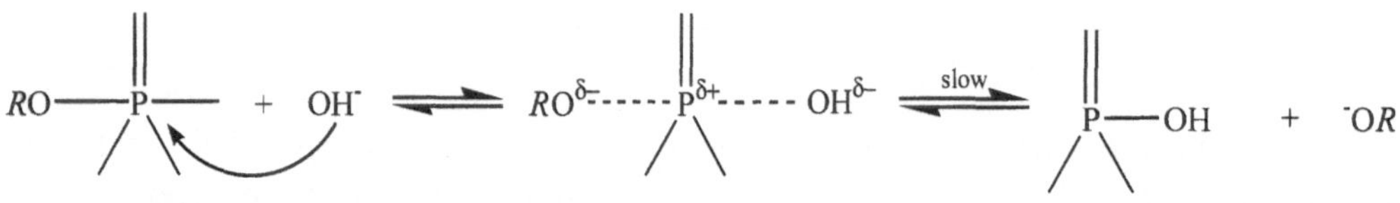

$$H_2O + {}^{-}OR \xrightarrow{\text{fast}} R\text{-}OH + OH^- \qquad (1)$$

Under hydrothermal conditions, the hydrolysis of di- and trialkylphosphates proceeds gradually, and the final product is phosphoric acid.

In the presence of Ca^{2+} ions, however, the second step proceeds probably in parallel with the hydrolysis according to Eq. (2):

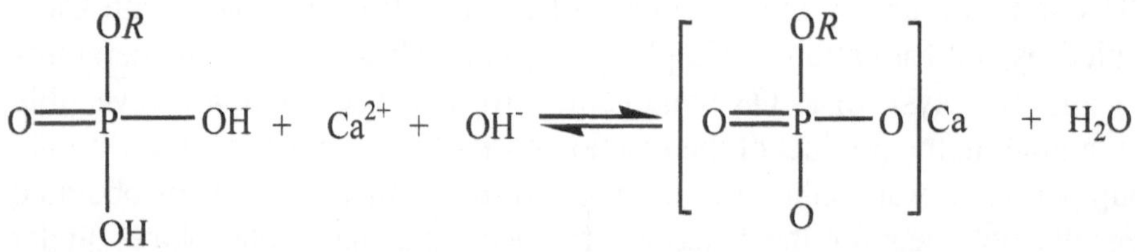

(2)

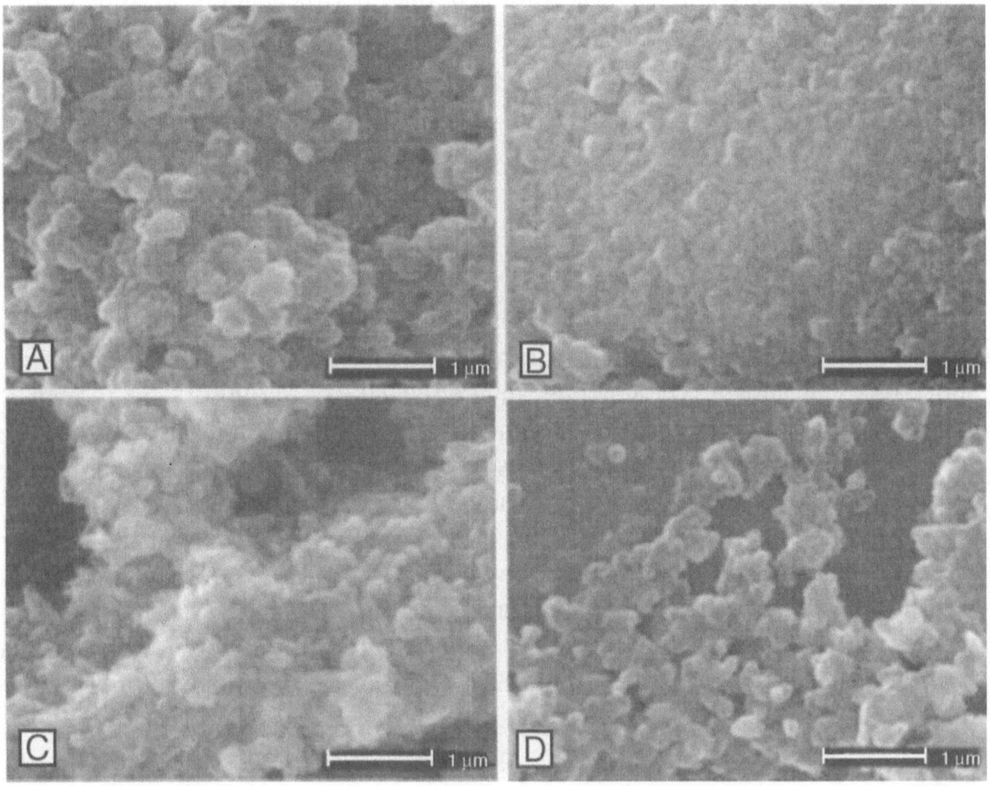

Fig. 5. SEM micrographs of hydroxyapatite particles prepared by hydrothermal synthesis at 200°C from A) methylphosphate, B) ethylphosphate, C) *i*-propylphosphate, D) *n*-butylphosphate

The existence of such or a similar intermediate product during hydrothermal synthesis can be inferred from the absorption of the CH_3 group in the IR spectrum (2970 cm^{-1}). Another fact that supports this hypothesis is the reaction product of the hydrothermal reaction of trialkylphosphates with Ca^{2+}, *i.e.* calcium hydrogenphosphate ($CaHPO_4$). It is assumed that the product is formed by hydrolysis of the last P-O*R* group of trialkylphosphate according to Eq. (3):

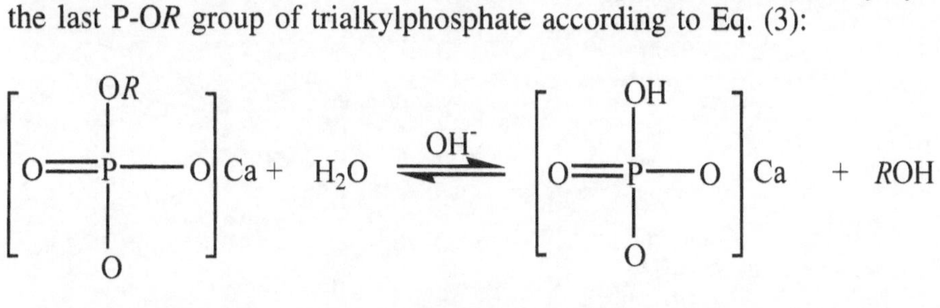

(3)

In the case of the hydrothermal reaction of mono- and dialkylphosphates, the transformation of $CaPO_3(OH)$ probably also proceeds according to the same equation.

The last step of the reaction mechanism is the hydrothermal transformation of $CaHPO_4$ into hydroxyapatite [18] according to Eq. (4):

$$3\,CaHPO_4 + 2Ca^{2+} + 4OH^- \longrightarrow Ca_5(PO_4)_3OH + 3\,H_2O \qquad (4)$$

A direct reaction of $CaPO_3(OH)$ according to the Eq. (5) cannot be excluded either:

$$3\left[\mathrm{O{=}P(OR)(O)\text{—}O}\right]\mathrm{Ca} + 4\,H_2O + 2\,Ca^{2+} + 4\,OH^- \longrightarrow Ca_5(PO_4)_3OH + 3\,R\mathrm{OH} + 3\,H_2O \qquad (5)$$

Hydroxyapatite coatings applied by the dip coating method

In Fig. 6, the surface of hydroxyapatite layers prepared by applying hydroxyapatite sols to Al_2O_3 ceramics by the dip coating method and subsequent sintering at 800–1100°C is shown. The coatings sintered at 800°C and 900°C are formed by hydroxyapatite particles about of 150 nm in size. Figure 6C shows the surface of a *HA* layer sintered at 1000°C; the layer is more compact than in the preceding case, and the particle size increased to 220 nm. The surfaces sintered at 1100°C were formed by sintered particles of 660 nm in size. An XRD analysis of the coatings revealed that hydroxyapatite transformed into β-$Ca_3(PO_4)_2$ during sintering at 1100°C. Hence, it follows that the transformation of nanocrystalline hydroxyapatite into β-$Ca_3(PO_4)_2$ took place at a temperature some 300°C lower than that established for hydroxyapatite ceramics with particle size around 5 μm [19, 20].

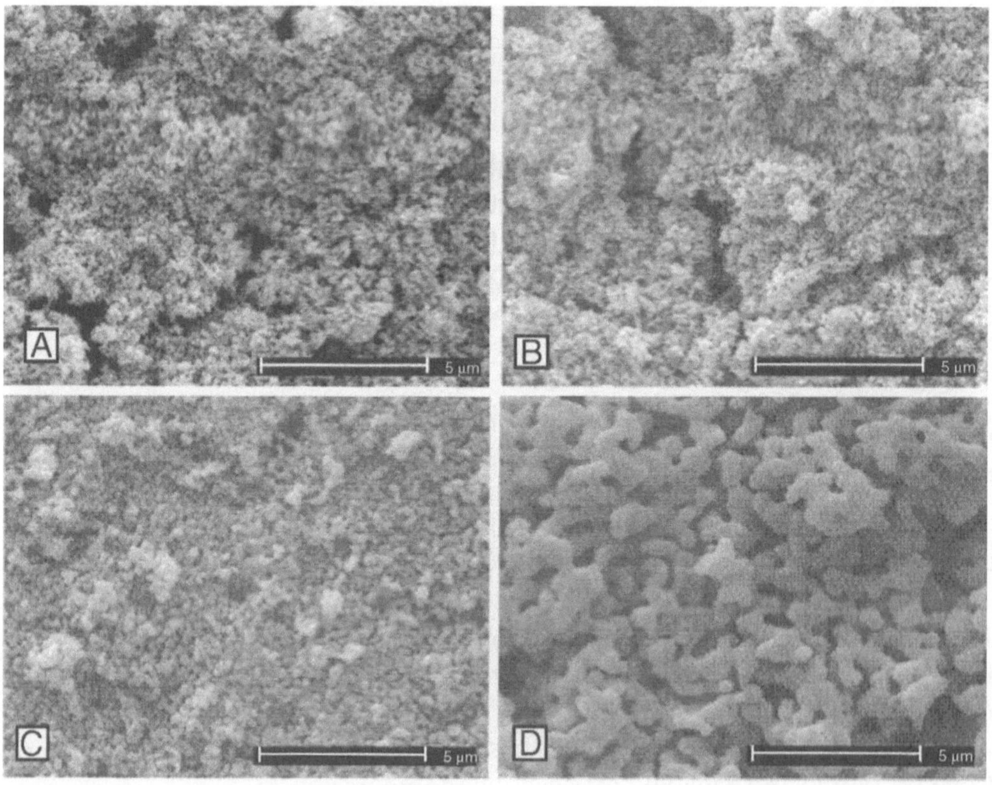

Fig. 6. SEM micrographs of hydroxyapatite coatings prepared by dip coating of hydroxyapatite sol on alumina substrate sintered at A) 800°C, B) 900°C, C) 1000°C, D) 1100°C

Conclusions

It was found that a hydrothermal reaction of mono- and dialkylphosphates with Ca^{2+} ions at a temperature of 160°C for 4 hours yielded nanocrystalline particles of hydroxyapatite of about 150 nm in size which are suitable for the preparation of colloidal suspensions and for applying bioactive layers by the dip coating method.

Experimental

Synthesis of hydroxyapatite

Hydroxyapatite was prepared by the reaction of calcium acetate with mono- and dialkylphosphate or trialkylphosphate (Fluka) at a molar Ca:P ratio of 1.67 in the presence of a base (*DEA*, NH_4OH). Mono- and dialkylphosphates (methyl, ethyl, *i*-propyl, *n*-butyl) were prepared by the reaction of P_2O_5 with the respective alcohol such that the resultant concentration of P in alkylphosphate was 2 mol/dm^3.

The synthesis of hydroxyapatite was conducted

i) under classical conditions: a solution of calcium acetate was added in drops into the reactor containing an alcoholic solution of the alkylphosphate. The *pH* value was maintained at 8–9 by addition of the base, and the reaction mixture was maintained at a temperature of 60–70°C for a period of 4 h.

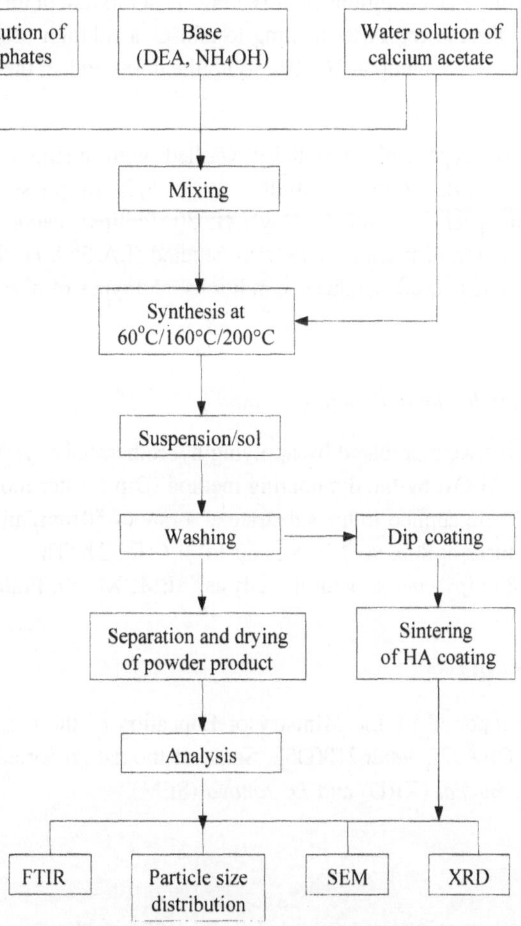

Fig. 7. Flow chart of nanocrystalline hydroxyapatite synthesis and analysis

Table 2. Conditions of hydroxyapatite synthesis

Synthesis type	Alkylphosphate	Base	Reaction time/h	Heat treatment
classical	mono- and dimethyl	NH_4OH	4	drying, calcination
classical	mono- and diethyl	NH_4OH	4	drying, calcination
classical	mono- and di-*i*-propyl	NH_4OH	4	drying, calcination
classical	mono- and di-*n*-butyl	NH_4OH	4	drying, calcination
HT/160°C	mono- and dimethyl	*DEA*	5	drying
HT/160°C	mono- and diethyl	*DEA*	5	drying
HT/160°C	mono- and di-*i*-propyl	*DEA*	5	drying
HT/160°C	mono- and di-*n*-butyl	*DEA*	5	drying
HT/200°C	mono- and dimethyl	*DEA*	5	drying
HT/200°C	mono- and diethyl	*DEA*	5	drying
HT/200°C	mono- and di-*i*-propyl	*DEA*	5	drying
HT/200°C	mono- and di-*n*-butyl	*DEA*	5	drying
HT/200°C	trimethyl	*DEA*	5	drying
HT/200°C	triethyl	*DEA*	5	drying

ii) under hydrothermal conditions (HT, see Fig. 7): the reactants were mixed together, poured into a hydrothermal reactor, and heated to 160°C (5 h, 16 MPa).

iii) under hydrothermal conditions (HT, see Fig. 7): a mixture of alkylphosphate and the base was poured into the reactor. After heating to 200°C, a solution of calcium acetate was added in doses by a pressure pump at 16 MPa. With the doses added, the mixture was heated to 200°C for 5 h.

The product was separated, repeatedly washed with distilled H_2O, and dried at 100°C. Synthesized *HA* in powder form was studied by RTG/X-ray phase analysis (Xpert, Philips), IR spectroscopy, scanning electron microscopy (XL 30, Philips); using laser diffraction, the particle sizes and the distribution of particle sizes were studied (LA-500, Horiba).

The types of reactants used, synthesis conditions, and types of heat treatment are summarized in Table 2.

Preparation of layers by the dip coating method

Hydroxyapatite layers were prepared by applying hydrothermally synthesized hydroxyapatite sol to ceramic substrates (Al_2O_3) by the dip coating method (Dip Master model 200, Chemat Technology Inc.). Three layers were applied to the substrate at a rate of 50 mm/min, and after drying they were sintered at temperatures of 800, 900, 1000, and 1100°C for 2 h. The coatings were studied by X-ray diffraction (Xpert, Philips) and structural analysis (SEM; XL 30, Philips).

Acknowledgements

This project was supported by the Ministry of Education of the Czech Republic under contracts COST 523.10 and CEZ:J22/98:262100002. Some of the data reported in this paper were obtained with the help of *A. Buchal* (XRD) and *D. Janova* (SEM).

References

[1] Lee BI, Samuels WD, Wang LQ, Exarhos GJ (1996) J Mat Res **11**: 134
[2] Cao Z, Lee BI, Samuels WD, Wang LQ, Exarhos GJ (1998) J Mat Res **13**: 1553

[3] Ali AF, Mustarelli P, Quartarone E, Magistris A (1999) J Mat Res **14**: 327
[4] Weng W, Baptista JL (1997) J Eur Ceram Soc **17**: 1151
[5] Kordas G, Trapalis CC (1997) J Sol-Gel Sci Tech **9**: 17
[6] Weng W, Baptista JL (1998) Biomater **19**: 125
[7] Gross KA, Chai CS, Kannangara GSK, Nissan BB (1998) J Mater Sci Mat Med **9**: 839
[8] Hattori T, Iwadate Y, Kato T (1989) J Mat Sci Let **8**: 305
[9] Liu HS, Chin TS, Lai LS, Chiu SY, Chung KH, Chang CS, Lui MT (1997) Ceram Inter **23**: 19
[10] Andres-Verges M, Fernandez-Gonzales C, Martinez-Gallego (18) J Eur Ceram Soc **18**: 1245
[11] Katsuki H, Furuta S (1999) J Am Ceram Soc **82**: 2257
[12] Hattori T, Iwadate Y (1990) J Am Ceram Soc **73**: 1803
[13] Lopez-Macipe A, Gomez-Morales J, Rodriguez-Clemente R (1998) Adv Mater **10**: 49
[14] Cihlar J, Castkova K (1998) Ceramics-Silikaty **42**:164
[15] Hattori T, Iwadate Y, Kato T (1988) Adv Ceram Mat **3**: 426
[16] Brendel T, Engel A, Russel C (1992) J Mater Sci Mat Med **7**: 175
[17] Szu S, Klein LC, Greenblatt M (1992) J Non-Cryst Solids **21**: 143
[18] Elliot JC (1994) Studies in Inorganic Chemistry 18: Structure and Chemistry of Apatites and Other Calcium Orthophosphates. Elsevier, Amsterdam, p 31
[19] Cihlar J, Buchal A, Trunec M (1999) J Mater Sci Mat Med **34**: 6123
[20] Cihlar J, Trunec M (1996) Biomater **17**: 1905

Received October 4, 2001. Accepted (revised) November 19, 2001

Preparation of Nanostructured Magnetic Films by the Plasma Jet Technique

Frantisek Fendrych[1,*], **Ludek Kraus**[1], **Oleksandr Chayka**[1], **Peter Lobotka**[2], **Ivo Vavra**[2], **Jan Tous**[1], **Vaclav Studnicka**[1], and **Zdenek Frait**[1]

[1] Institute of Physics, ASCR, CZ-18221 Praha 8, Czech Republic
[2] Institute of Electrical Engineering, SAS, SK-84239 Bratislava, Slovak Republic

Summary. Magnetic films were prepared by the plasma jet technique from Fe, mumetal, and Fe/Hf or Fe/Ta nozzles. Two different plasma jet systems with different vacuum pumps were used to compare the quality of the produced films. The films prepared from a Fe nozzle in the two different equipments shows that oxygen in the residual atmosphere of the low vacuum reactor leads mainly to the formation of iron oxides. The Fe and mumetal films prepared in the high vacuum system contain only a very small amount of oxygen, as proved by chemical analysis and ferromagnetic resonance. The mumetal film, moreover, shows good soft magnetic properties and low magnetic damping. For the reactive plasma jet deposition of nanogranular Fe–Hf–O and Fe–Ta–O films, the low vacuum system was used. The films with higher oxygen content exhibit tunneling-type conductivity. In some films, superparamagnetic behaviour and spin-dependent tunneling magnetoresistance were observed.

Keywords. Plasma jet; Magnetic films; Nanogranular materials; Tunneling magnetoresistance.

Introduction

Magnetic films are nowadays very important, both from technical application and basic research points of view. A large variety of methods is being used to produce magnetic films. For the new applications, however, new magnetic materials and also new deposition techniques are required. The plasma jet technique described below has only recently been used for the preparation of magnetic films [1–4].

The plasma jet technique was developed in the Institute of Physics, Czechoslovak Academy of Sciences [5]. The principle of the method is schematically shown in Fig. 1. The material to be deposited originates from the nozzle, which is sputtered by the ions produced in the hollow cathode discharge generated inside the nozzle. In the basic configuration the plasma jet reactor consists of two electrodes: the grounded walls of the reactor (anode) and the water-cooled cylindrical nozzle (cathode) which is connected to the DC or radio frequency power source. The substrate is placed on the electrode, which can be grounded or biased with respect to the ground potential. The working gas (or mixture of gases) enters

* Corresponding author. E-mail: fendrych@fzu.cz

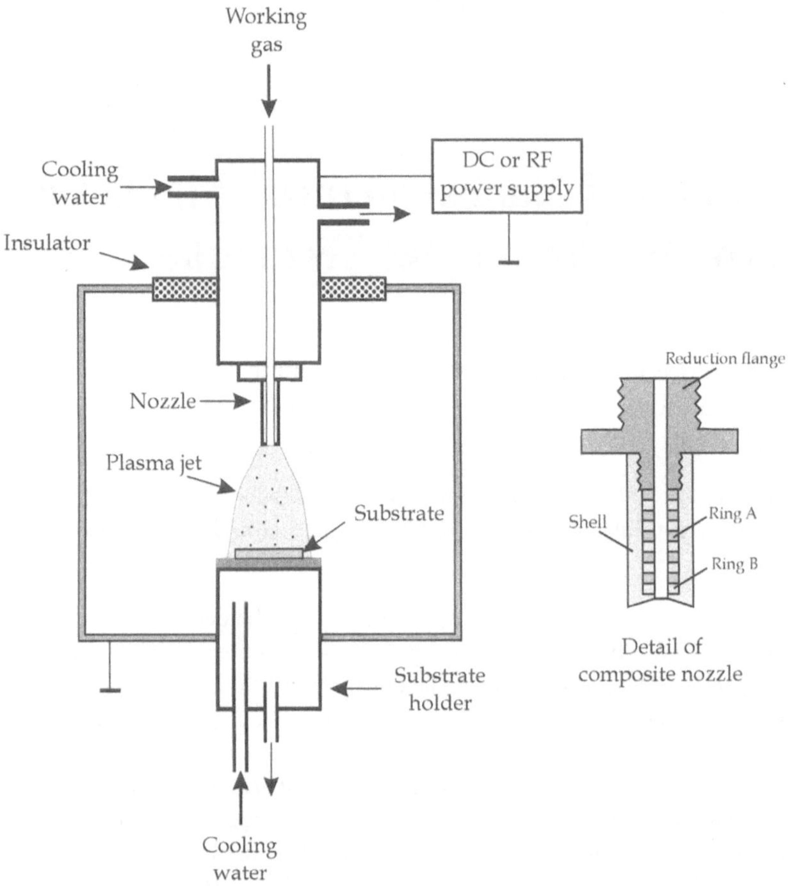

Fig. 1. Schematic view of plasma jet reactor

the reactor through the nozzle and is continuously pumped by the vacuum system. The pressure in the reactor chamber is maintained at several tens of Pa. The flow of working gas and the pumping speed can be adjusted so that the velocity of gas at the outlet of the nozzle is supersonic. If the power fed to the nozzle exceeds a certain limit (for example, 20 W), an intensive hollow cathode discharge is generated inside the nozzle. Due to the bombardment by the ions produced in the discharge, the material of nozzle mouth is sputtered and enters the plasma. The working gas forces the discharge supersonically out of the nozzle, and a well-defined plasma jet superimposed on the primary plasma channel is formed. The atoms, ions, and clusters are transported to the substrate placed in the stream of plasma, and a film consisting of the nozzle material is deposited. In case of reactive plasma jet deposition a gas reacting with the nozzle material is added to the working gas, and chemical compounds can be then produced. More details on the plasma jet reactor can be found in Refs. [6, 7]. The advantage of the plasma jet technique with respect to the commonly used magnetron sputtering is the high deposition rate (of the order 1 μm/h) which can be achieved even for soft magnetic materials.

The plasma jet method has been successfully used for the preparation of various materials. Typical examples are the depositions of germanium nitride [8], cooper nitride [9], and titanium nitride [10] thin films of defined stoichiometry, composite SiGe [11], *a*-Si:H [12], hard amorphous carbon nitride coatings [13, 14], *etc.*

Results and Discussion

Fe films

The films prepared from a single Fe nozzle in the two different plasma jet equipments differ substantially. Chemical analysis showed that the films prepared in the low-vacuum reactor contained large amounts of oxygen. The film composition as a function of position, measured along the line passing the center of plasma channel, is shown in Fig. 2. Three films prepared under identical conditions but with different working gas are compared. As can be seen, the oxygen content does not depend on the purity of the working gas. That means that the oxygen in the film originates from the residual atmosphere in the reactor. Its amount could not be reduced below 30 at.% even if the reactor was pumped for one week before the deposition. To reduce the oxygen content in the films, some H_2 was added to the Ar. This leads to a substantial reduction of oxygen in a small circular spot around the plasma channel axis. Further from the substrate center, even higher oxidation of the film is observed. The quality of the film within the spot is rather poor with very small adhesion. X-Ray diffractions of these films indicate the presence of Fe oxides. Only weak traces of metallic Fe can be found in the films prepared in an Ar/H_2 atmosphere. No FMR signal can be observed from films prepared with Ar only; the film prepared in Ar/H_2 has not been measured because of its poor quality.

The films deposited in the high-vacuum reactor from the Fe nozzle on Si substrates show metallic luster and good adhesion to the substrate. Chemical analysis reveals only little contamination by oxygen (less than 6 at.%). The FMR measured with the magnetic field parallel to the film plane is shown in Fig. 3. In the

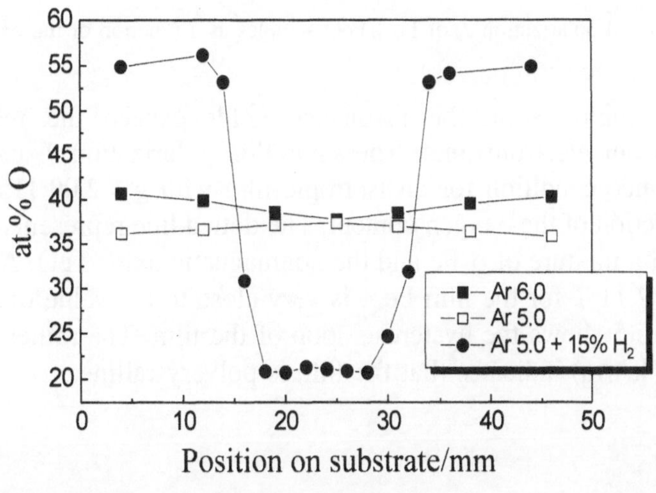

Fig. 2. Oxygen content in Fe-films prepared in the low-vacuum reactor with different working gas atmosphere

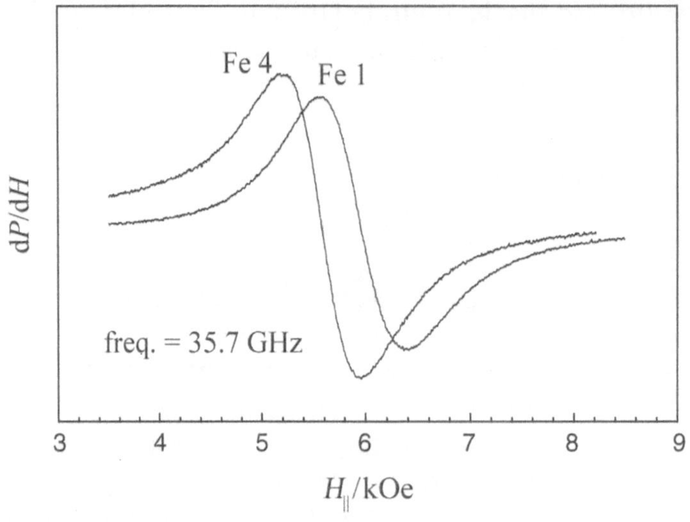

Fig. 3. Ferromagnetic resonance on Fe films deposited in the high-vacuum reactor

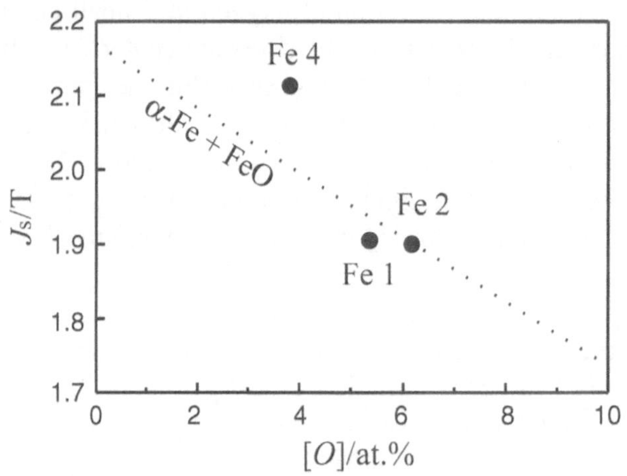

Fig. 4. Saturation polarization J_s of Fe + FeO samples as a function of the oxygen content

perpendicular configuration the resonance fields exceed the maximum field available with our electromagnet. The saturation polarization J_s calculated from *Kittel*'s resonance condition for an isotropic film with $g = 2.09$ [15] is shown in Fig. 4 as a function of the oxygen content. The dotted line represents the saturation polarization of a mixture of α-Fe and the nonmagnetic oxide FeO. As can be seen, the value $J_s = 2.11$ T for the film Fe 4 is very close to the value of 2.17 T for the bulk Fe. Figure 5 shows the hysteresis loop of the film. The rather high coercive force (about 3 kA/m) indicates that the film is polycrystalline.

Mumetal films

The mumetal films were prepared in the high-vacuum apparatus from the PY 79M nozzle. The films were deposited either on glass or Si substrates.

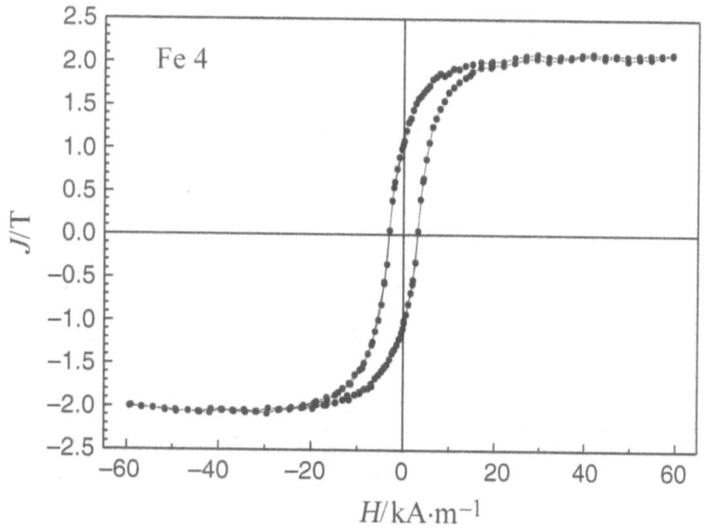

Fig. 5. Hysteresis loop of Fe film deposited in the high-vacuum plasma jet reactor

Electron probe microanalysis showed that the film stoichiometry did not differ by more than 2 at.% from the nozzle composition and that the increase of oxygen content with respect to the nozzle material was less than 0.4 at.% (in the bulk material, 1.8 at.% O were detected). X-ray diffraction pattern of the film on Pyrex glass shows diffraction peaks typical for an fcc-cubic structure with a lattice constant of $a = 3.554$ Å.

The results of ferromagnetic resonance measurements at 35.7 GHz with the magnetic field applied parallel and perpendicular to the film are shown in Fig. 6.

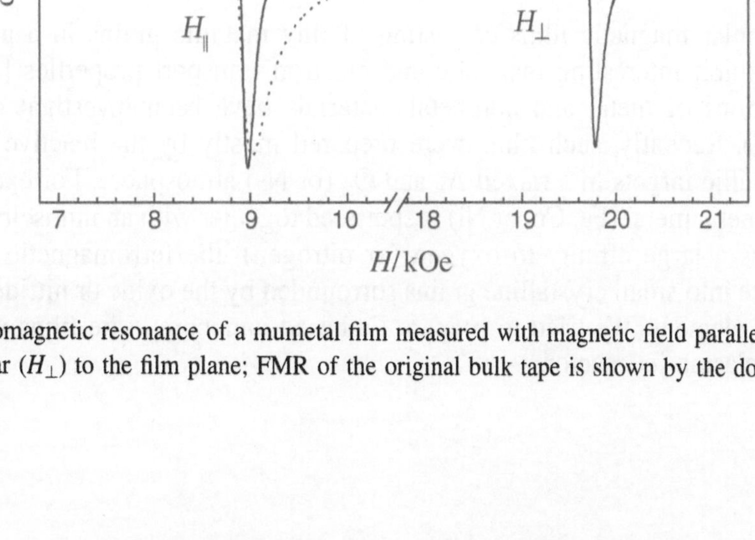

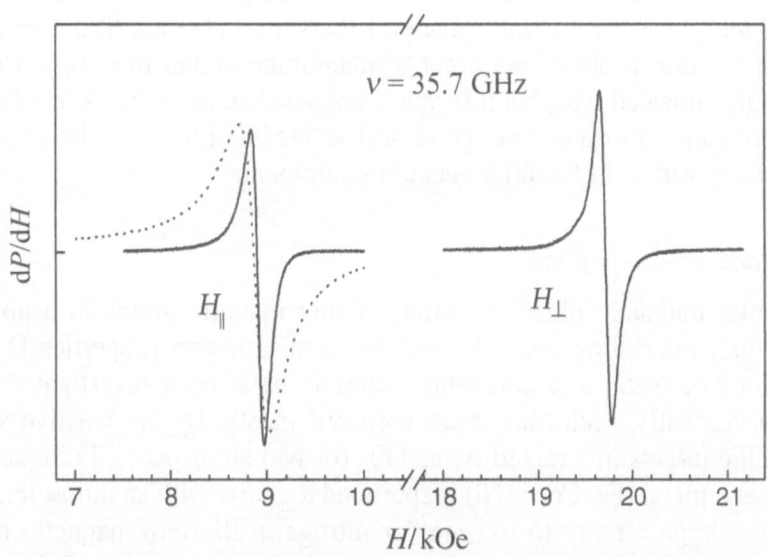

Fig. 6. Ferromagnetic resonance of a mumetal film measured with magnetic field parallel ($H_{\parallel}$) and perpendicular ($H_{\perp}$) to the film plane; FMR of the original bulk tape is shown by the dotted curve

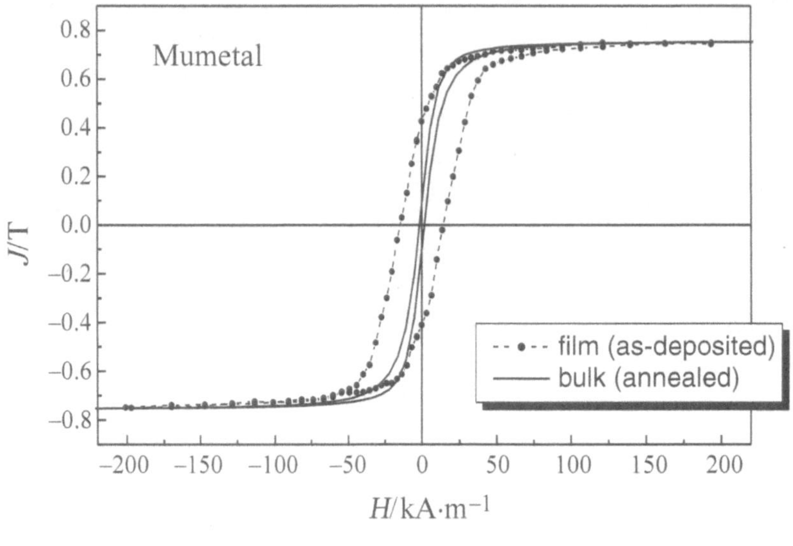

Fig. 7. Hysteresis loops of the mumetal film and the bulk mumetal tape

For comparison, the parallel resonance of the original PY 79M tape is also shown. From *Kittel*'s resonance condition for an isotropic film, the spectroscopic splitting factor $g = 2.108$ and the saturation polarization $J_s = 0.77\,T$ are obtained. This J_s value is identical with J_s of the original bulk alloy obtained from the hysteresis loop measurement-another proof that only very little oxygen is present in the μ-metal film. The FMR linewidth measurements at two different microwave frequencies (35.7 and 69 GHz) yield the intrinsic *Landau-Lifshitz* relaxation constant as 1.04×10^8 rad/s, in good agreement with values for 80/20 permalloy ($\sim 8 \times 10^7$ rad/s) films [16]. The low value of relaxation constant indicates a good quality of the plasma deposited film.

In Fig. 7, the hysteresis loop of the film is compared with the loop of a bulk tape (0.05 mm thick, 3 mm wide) made of the same material. The coercive force 14 A/m of the film is about one order of magnitude higher than $H_c = 1.6$ A/m of the optimally annealed tape, but it is much smaller than $H_c = 365$ A/m of the rolled tape. This result illustrates that good soft magnetic films can be produced by plasma jet deposition in the high-vacuum equipment.

Fe–Hf–O and Fe–Ta–O films

Nanogranular magnetic films consisting of tiny metallic grains in a nonmetallic matrix exhibit interesting magnetic and electron transport properties [17]. Many combinations of metal and nonmetal materials have been investigated (see *e.g.* Ref. [18]). Recently, such films were prepared mostly by the reactive sputtering from metallic targets in a mixed Ar and O_2 (or N_2) atmosphere. For example, if a ferromagnetic metal (Fe, Co, or Ni) is sputtered together with an inmiscible element which has a large affinity to oxygen (or nitrogen), the ferromagnetic metal can precipitate into small crystalline grains surrounded by the oxide or nitride matrix of the other element. We tried to prepare Fe-based nanocomposite films in the low-vacuum plasma jet reactor.

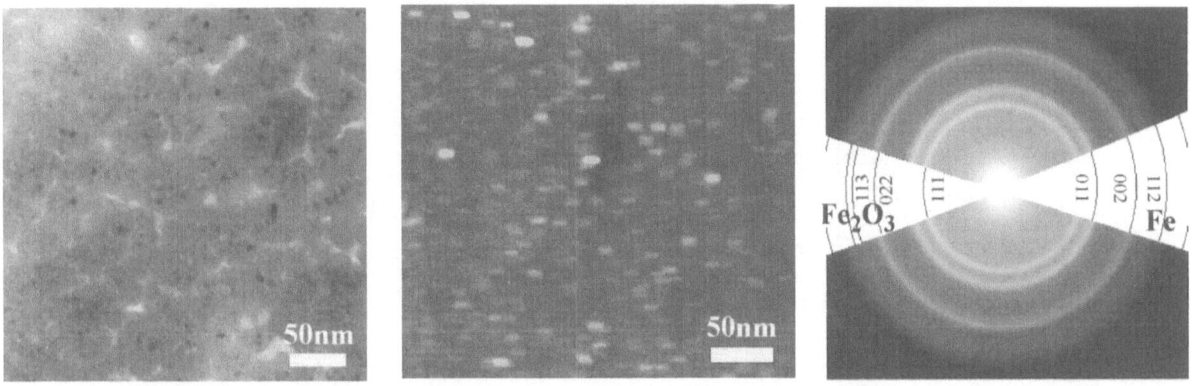

Fig. 8. TEM micrographs and electron diffraction of a nanogranular $Fe_{33.7}Ta_{9.5}O_{56.8}$ film; a) bright field image, b) dark field image, c) selective area diffraction

Because the partial pressure of oxygen in the low-vacuum reactor is rather high, O_2 need not be added to Ar to obtain the oxides of Hf or Ta. Moreover, sometimes hydrogen had to be added to Ar to prevent the oxidation even of the ferromagnetic metal. The Fe–Hf–O and Fe–Ta–O films prepared from the composite nozzles were investigated systematically [1–4]. Some of them really exhibit a nanogranular structure as shown by electron transmission microscopy (Fig. 8a,b). Bcc–Fe grains with a grain size of few nm can be found in an amorphous matrix. The electron diffraction revealed also some traces of crystalline Fe_3O_3 (Fig. 8c). It means that the surface of the bcc–Fe nanocrystals is probably oxidized. X-Ray diffraction of such films shows only a broad weak diffraction peak near the [110] reflection of bcc–Fe on an amorphous background. This fact is in correlation with TEM investigations because of the lower sensitivity of X-ray diffraction analysis.

Magnetic and electric properties of the films depend on the content of oxygen and on the Hf/Fe (or Ta/Fe) ratio. Ferromagnetic resonance measurements on Fe–Ta–O films revealed superparamagnetic behaviour, characterized by broad resonance peaks near the paramagnetic resonance field, for the compositions with a Ta/Fe ratio close to 0.25 [3].

The electrical resistivity ρ and its temperature dependence are very sensitive to the oxygen concentration. A large increase of ρ (by about 6 orders of magnitude) is observed for a content of above 50 at.% O. The temperature dependence of resistivity changes from the typical metallic behaviour at low O content to $\ln\rho \propto T^{1/2}$ for high O concentration (Fig. 9). Such a dependence is characteristic for the electron tunneling in metal/insulator granular systems [19]. The tunneling type of conductivity can be also proved by the dependence of differential conductance, $G = dI/dV$, on the bias voltage V. An example for a nanogranular Fe–Ta–O film is shown in Fig. 10. The fit of experimental data to the theoretical dependence $G \propto V^{1/2}$, typical for the inelastic tunneling of electrons through the non-metallic barrier [20], is shown by the dotted line.

In the films with tunneling type of conductivity and superparamagnetic behaviour, a high magnetoresistance (MR) was observed. An example can be seen in Fig. 11, where the MR of a film prepared from a Fe/Hf nozzle with a stainless

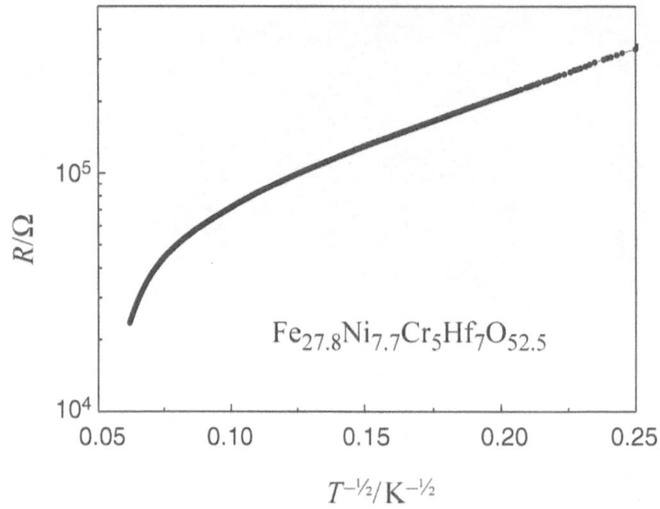

Fig. 9. Temperature dependence of resistivity of a nanogranular $Fe_{27.8}Ni_{7.7}Cr_5Hf_7O_{52.5}$ film

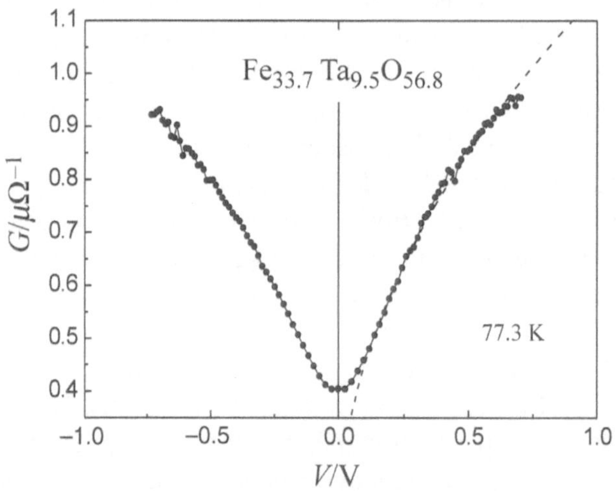

Fig. 10. Dependence of differential conductance $G = dI/dV$ on the bias voltage measured for a nanogranular Fe–Ta–O film at liquid nitrogen temperature

steel shell tube is shown. The magnetoresistance is isotropic in the film plane and follows well the theoretical dependence $\Delta\rho/\rho \sim -M^2$ [21]. The *m-H* curves, measured with the SQUID magnetometer at three different temperatures, are shown in the lower part of Fig. 11. In spite of similar magnetizing curves observed at 77.3 K and 300 K, the magnitudes of MR measured at these temperatures are very different. The enhancement of MR at lower temperature can be explained by higher-order spin-dependent tunneling [22]. We have tested this explanation by fitting the experimental curve *R*(V) measured at 77.3 K by the formula suggested in Ref. [22]. The result is shown in Fig. 12; a good agreement with the theoretical curve is evident. The MR enhancement is achieved due to the fact that the

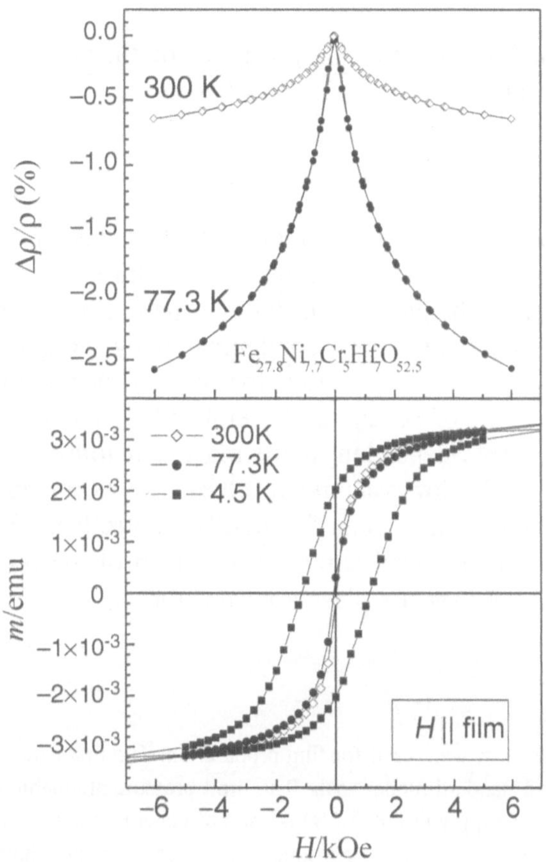

Fig. 11. Magnetoresistance (top) and the magnetizing curves (bottom) measured at different temperatures in a nanogranular $Fe_{27.8}Ni_{7.7}Cr_5Hf_7O_{52.5}$ film

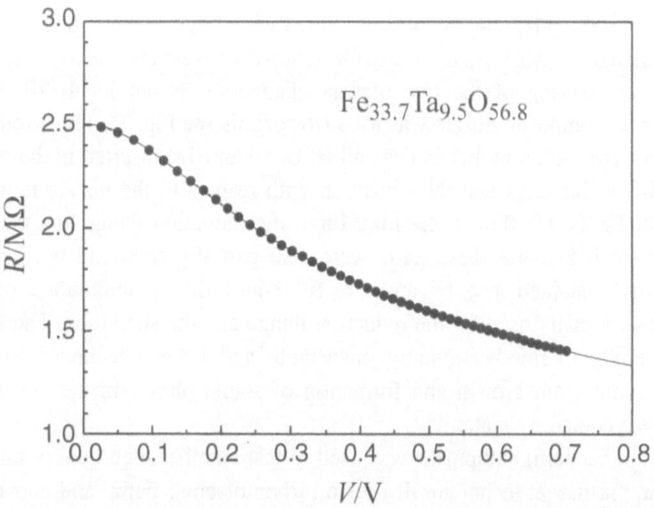

Fig. 12. Fit of the experimental dependence of the Fe–Ta–O film resistance R on the bias voltage V measured at 77.3 K; the fitting formula $R(V) = a(1 + bV^2)^{-C}$ was proposed by Mitani et al. in Ref. [22] assuming higher-order spin-dependent tunneling in granular systems when several electrons placed at neighboring granules tunnel simultaneously between them

probability of the higher-order tunneling is given by the product of probabilities of single tunneling events.

Conclusions

For the first time the supersonic plasma jet method has been used for deposition of magnetic films. It was proved that the quality of film sensitively depends on the vacuum system used for the pumping of the plasma reactor. The films deposited in the low vacuum reactor are highly contaminated by oxygen from the residual atmosphere. This system has been used for reactive deposition of nanogranular magnetic films consisting of Fe crystallites embedded in an oxide matrix. Some of these films exhibited spin-dependent tunneling magnetoresistance. The quality of the films prepared in the low vacuum reactor is, however, rather poor, and the chemical composition of films can be hardly controlled. To get high quality magnetic films with low (or well controlled) content of oxygen, the high-vacuum technology must be used for the plasma jet reactor.

Experimental

Two different plasma reactors were used for film preparation. The first (low vacuum) system uses a Roots vacuum pump and standard rubber seals. The limit pressure attainable in this reactor is about 5×10^{-2} Pa. An RF power supply (13.56 MHz) is used to generate the hollow cathode discharge in the nozzle. The single Fe nozzle (tube 3 cm long with inner and outer diameters of 3 and 6 mm, respectively) was used to test the system capability. Because the films prepared in this reactor are highly contaminated by the oxygen from the residual atmosphere, H_2 was added to the working gas (Ar) in order to reduce the oxygen content in some films. The low vacuum system was also used for the preparation of nanogranular magnetic films by reactive deposition. Deposition from two independent nozzles (Fe in combination with Hf or Ta) at 90° and 45° with respect to the substrate plane was attempted. The idea was to control the film composition by variation of the Fe and Hf (Ta) nozzle's distance from the substrate, by different flow rates of the working gas, or by a difference in the RF power supplied to the two nozzles. These films, however, were chemically very inhomogeneous, probably because the mixing of the two plasma channels was not ideal. To improve the film homogeneity, a single composite nozzle was used (for details see Fig. 1). The nozzle was composed of rings made of the two different metals (Fe and Hf or Fe and Ta) inserted in the outer (shell) tube. The number of Hf (or Ta) rings and their position with respect to the nozzle mouth were used to control the ratio of Hf/Fe (Ta/Fe) in the film. First, the reduction flange and the shell tube were made of stainless steel. Because these parts were also partially sputtered by the hollow cathode discharge, the films contained also Ni and Cr. To reduce the contamination of films by other elements, pure Fe was finally used for the reduction flange and the shell tube. These nanocomposite materials are generally thermodynamically metastable and have a tendency to relax at higher temperatures. To avoid grain growth and formation of stable phases in the developing films, the substrate holder was cooled by water.

The second (high vacuum) equipment was used to test whether high quality magnetic films can be prepared by the plasma jet technique. It uses a turbomolecular pump and copper vacuum seals. The limit pressure in this reactor is about 10^{-6} Pa. The substrate holder is not cooled. The DC hollow cathode discharge with plasma confined by a magnetic field was used for the deposition of Fe and μ-metal films. The same Fe nozzle as in the low vacuum system was used for Fe film deposition. The μ-metal films were prepared from a nozzle made of $Ni_{73}Fe_{15}Cu_7Mo_4Mn_1$ alloy. The mother alloy (commercial mark: PY 79M) was kindly supplied by Kovohute Rokycany, a.s.

The chemical composition of the films was studied by electron probe microanalysis on a JEOL Superprobe JXA-733 device equipped with an X-ray microanalyzer Kevex Delta class V. Two methods of analysis were used:

1. EDAX (energy dispersive analysis of characteristic X-ray radiation) using a multichannel semiconductor detector with a special beryllium window for light elements (up from boron).
2. WDAX (wave dispersive analysis of X-rays) using adjustable crystal spectrometers. A special multilayer crystal Ni–C (spacing: $2d = 84$ Å) was used as an X-ray diffraction mirror for the direct measurement of light elements (*e.g.* O, N). Pure samples of CuO and cubic BN were used as standards for O and N measurement, respectively.

First, the EDAX method was applied for the identification of chemical elements in the film and a rough estimation of their concentrations. WDAX was then utilized for the precise quantitative analysis of chemical elements identified by EDAX. In this way, an accuracy of chemical composition of about ± 0.1 wt.% was achieved.

The X-ray diffraction structure analysis was performed by means of a *Bragg-Brentano* focusation arrangement using CuK_α radiation. The magnetic properties were studied at room temperature by a computer controlled quasistatic hysteresis loop tracer. The saturation magnetization, g-factor, and magnetic relaxation parameter of the deposited films were determined from ferromagnetic resonance (FMR) at 36 and 69 GHz with the magnetic field applied parallel and perpendicular to the film plane. For the measurement of magnetization curves in nanogranular films the SQUID magnetometer (Quantum Design MPMS-5S) was used.

The resistance of low resistive samples (less than 200 MΩ) was measured by the four-terminal method using a conventional digital multimeter. For high resistive samples, the two-terminal method and an electrometer were used. For the resistance measurement the films were deposited on glass substrates. The electric contacts were prepared by electron gun deposition of Ag films either below or on the top of the investigated film. The electrical leads were glued to the Ag films by silver paint. The temperature dependence of resistivity was examined in the range of 4.2–300 K. The magnetoresistance at 77 K and room temperature was measured in an electromagnet with its magnetic field (up to 6 kOe) applied parallel or perpendicular to the film plane.

Acknowledgements

This work has been supported by the COST-523 European Concerted Action on Nanostructured Materials and by the Grant Agency of the Czech Academy of Sciences under Project No. A1010204. The paper was presented at the COST-523 Mid Term Meeting in Limerick, Ireland, October 4–6, 2001.

References

[1] Soyka V, Kraus L, Fendrych F, Sicha M, Hubicka Z, Frait Z, Jastrabik L (1999) 26th Int Conf on Metallurgical Coatings and Thin Films, April 12–15, 1999, San Diego, California, USA, Book of Abstracts, p 85

[2] Soyka V, Kraus L, Frait Z, Sicha M, Jastrabik L (2000) Acta Physica Polonica A **97**: 515

[3] Kraus L, Chayka O, Tous J, Fendrych F, Pirota KR, Sicha M, Jastrabik L (2001) Int Conf on Magnetism, August 6–11, 2000, Recife, Brazil; J Magn Magn Mater **226–230**: 669

[4] Kraus L, Chayka O, Fendrych F, Frait Z, Sicha M, Tous J (2000) 16th Int Conf on Magnetic Films and Surfaces, August 14–18, Natal, Brazil, Book of Abstracts, paper 15P24

[5] Bardos L, Vu NQ (1989) Czech J Phys B **39**: 731

[6] Novak M, Sicha M, Kapicka V, Jastrabik L, Soukup L, Hubicka Z, Klima M, Slavicek P, Brablec A (1997) Journal de Physique IV Coloque Suppl C4-331

[7] Hubicka Z, Pribil G, Soukup RJ, Ianno NJ (2002) Surface and Coatings Technology (submitted)
[8] Soukup L, Perina V, Jastrabik L, Sicha M, Pokorny P, Soukup RJ, Novak M, Zemek J (1996) Surface and Coatings Technology **78**: 280
[9] Fendrych F, Soukup L, Jastrabik L, Sicha M, Hubicka Z, Chvostova D, Tarasenko A, Studnicka V, Wagner T (1999) Diamond and Related Materials **8**: 1715
[10] Barankova H, Bardos L, Berg S (1995) J Electrochem Soc **142**: 883
[11] Sicha M, Hubicka Z, Soukup L, Jastrabik L, Cada M, Spatenka P (2001) Surface and Coatings Technology **148**: 199
[12] Pribil G, Hubicka Z, Soukup RJ, Ianno NJ (2001) J Vacuum Science Technol **A19(4)**: 1571
[13] Fendrych F, Pajasova L, Wagner T, Jastrabik L, Chvostova D, Soukup L, Rusnak K (1999) Diamond and Related Materials **8**: 1711
[14] Hubicka Z, Sicha M, Pajasova L, Soukup L, Jastrabik L, Chvostova D, Wagner T (2001) Surface and Coatings Technology **142–144**: 681
[15] Frait Z, Gemperle R (1971) Journ de Physique **32**: C1-541
[16] Frait Z, Fraitova D (1998) In: Wigen P, Baryachtar V, Lesnik N (eds) Frontiers in Magnetism of Reduced Dimension Systems, NATO ASI, Series 3, Kluver, Dordrecht, vol 49, pp 121–152
[17] Mitani S, Fujimori H, Ohnuma S (1997) J Magn Magn Mater **165**: 141
[18] Mitani S et al. (1999) J Magn Magn Mater **198, 199**: 179
[19] Sheng P, Abeles B, Arie Y (1973) Phys Rev Lett **31**: 44
[20] Altshuler BL, Aronov AG (1979) Sov Phys JETP **77**: 2028
[21] Inoue J, Maekawa S (1996) Phys Rev B **53**: R11927
[22] Mitani S, Takahashi S, Takanashi K, Yakushiji K, Maekawa S, Fujimori H (1998) Phys Rev Lett **81**: 2799

Received October 5, 2001. Accepted November 22, 2001

Growth of Anodic Films on Compound Semiconductor Electrodes: InP in Aqueous $(NH_4)_2S$

Denis N. Buckley[1,*], **Elizabeth Harvey**[1], and **Sung-Nee G. Chu**[2]

[1] Department of Physics and Materials and Surface Science Institute, University of Limerick, Ireland

[2] Agere Systems, Murray Hill, NJ, USA

Summary. Film formation on compound semiconductors under anodic conditions is discussed. The surface properties of InP electrodes were examined following anodization in an $(NH_4)_2S$ electrolyte. The observation of a current peak in the cyclic voltammetric curve was attributed to selective etching of the substrate and a film formation process. AFM images of samples anodized in the sulfide solution revealed surface pitting. Thicker films formed at higher potentials exhibited extensive cracking as observed by optical and electron microscopy, and this was explicitly demonstrated to occur *ex situ* rather than during the electrochemical treatment. The composition of the thick film was identified as In_2S_3 by EDX and XPS. The measured film thickness varies linearly with the charge passed, and comparison between experimental thickness measurements and theoretical estimates for the thickness indicate a porosity of over 70%. Cracking is attributed to shrinkage during drying of the highly porous film and does not necessarily imply stress in the wet film as grown. During the growth of the thick porous film, spontaneous current oscillations have been observed. The frequency of oscillation was found to be proportional to the current density, regardless of whether the measurements were carried out during a potential sweep or at constant potential. Thus, the charge passed per oscillation remained constant. A characteristic value of approximately $0.3\,C \cdot cm^{-2}$ was measured under potential sweep conditions, and a similar value was obtained at constant potential.

Keywords. InP; Anodic films; Oscillations; Cyclic voltammetry; Electron microscopy.

Introduction

Group III–V semiconductors are widely used for optoelectronic devices and for high power and high speed electronic devices. However, passivation of the semiconductor is required to overcome some of the device problems which are associated with surface states, such as catastrophic optical damage in lasers [1] and low current gain in heterojunction bipolar transistors (HBTs) [2]. Study of the anodic growth of films on III–V semiconductors such as InP [3] and GaAs [4] has stemmed from their potential use as insulating layers in metal-insulator

* Corresponding author. E-mail: noel.buckley@ul.ie

semiconductor (MIS) devices with better electrical results being achieved when a passivation step is carried out prior to growth of the anodic film [5, 6]. Photoenhanced oxidation of GaN has been found to result in an increase in photoluminescence intensity, thus indicating the passivating nature of the surface oxide formed in this case [7]. Metal-oxide semiconductor (MOS) structures using photo-electrochemically formed oxide layers on GaN have been fabricated [8], and interesting features such as cracking of the oxide film formed have also been reported for the case of GaN [7].

There is considerable interest in the role of sulfur as a passivating species [9–11]. Sulfur treatment has been shown to considerably improve surface properties of III–V semiconductors [12–14]. In particular, anodic films grown in sulfur-containing electrolytes appear to result in surfaces which are more resistant to reoxidation [15, 16]. However, cracking of the anodically grown surface film on a p-InP electrode in a sulfur-containing electrolyte has been reported [17].

The observation of current oscillations during the anodization of CdTe [18], HgTe [19], and CdHgTe [20] in sulfide solutions has been reported. Anodic oscillations have also been observed in n-GaAs under conditions of strong illumination in a borax solution [21] and in both p-type [22] and n-type [23] Si, mainly in fluorine-containing electrolytes. More recently, potential oscillations have been noticed during galvanostatic anodization of n-InP in HCl [24]. Cases of oscillatory behaviour have been reported for other semiconductor/electrolyte systems [25] and for many metal/electrolyte systems [26–28].

Thus, the nature of anodic processes on compound semiconductors, the formation of films, their structure and composition, and the relationship to growth conditions are important, both from a fundamental and a technological point of view. It seems worthwhile to gain deeper insight into the factors governing the electrochemical reactions, the film growth kinetics and behaviour, the morphology, and other properties of the films formed in order to understand the mechanism of dissolution, film growth, and oscillatory behaviour of compound semiconductors.

Due to the above mentioned interest in interfaces between compound semiconductors and sulfur-containing electrolytes we have chosen to use InP electrodes in aqueous sulfide solutions as a model system to investigate their behaviour [29, 30]. In this paper we report on the results of an investigation of the growth of anodic films in this system. We elucidate the nature of previously reported cracking of these films and communicate observations of electrochemical oscillations.

Results and Discussion

Film growth at lower potentials

Figure 1 shows cyclic voltammograms of an InP electrode in a $3\,\mathrm{mol\cdot dm^{-3}}$ $(NH_4)_2S$ electrolyte. The potential was scanned at a rate of $10\,\mathrm{mV\cdot s^{-1}}$ between an initial value of 0.0 V and upper potentials (E_U) of 0.785 and 0.88 V, respectively. When E_U was less then the peak potential E_P, the value of the current on the return scan was roughly similar to the corresponding current on the forward scan (Fig. 1a). The current-voltage characteristics obtained for $E_U = 0.88$ V (*i.e.* above the peak potential) are shown in Fig. 1b. In this case, the current density on the cathodic

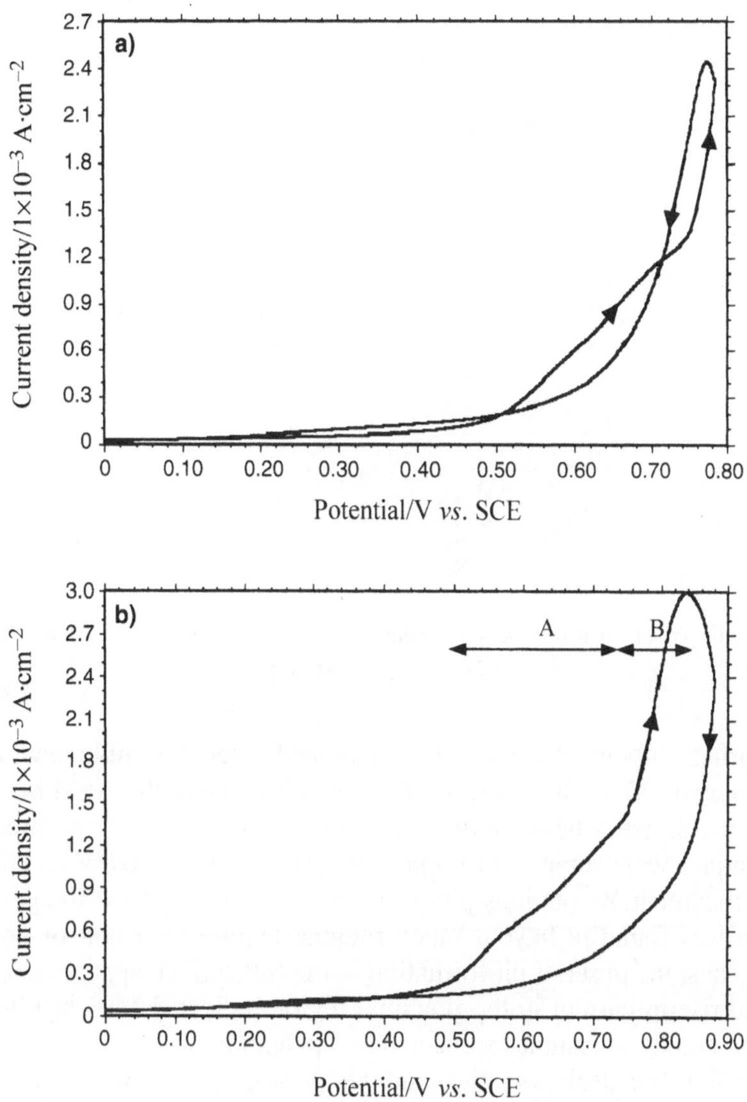

Fig. 1. Cyclic voltammograms of InP in 3 mol · dm^{-3} $(NH_4)_2S$ at a scan rate of 10 mV · s^{-1} between 0.0 V (SCE) and a) $E_U = 0.785$ V (SCE) b) $E_U = 0.88$ V (SCE)

scan was considerably lower than on the anodic scan. Such results suggest that the peak in current density corresponds to passivation of the surface by a deposited film which also inhibits current flow on the cathodic scan.

Atomic force microscopy (AFM) was used to investigate the topography of the InP electrodes following different anodization procedures. Figure 2 shows a 500 nm × 500 nm AFM image of a sample that was subjected to a linear sweep from 0.0 V to a value of $E_U = 0.785$ V. On careful inspection it can be seen that small holes or pits are dispersed over the surface of the sample. Larger areas of the same electrode were imaged, and these pits were found to exist over the whole surface. This indicates that selective etching of the InP substrate occurs in region B of Fig. 1b. If the potential was scanned back to 0.0 V, an increase in the pit density was observed. Thus, selective etching of the InP electrode also occurs on the return scan.

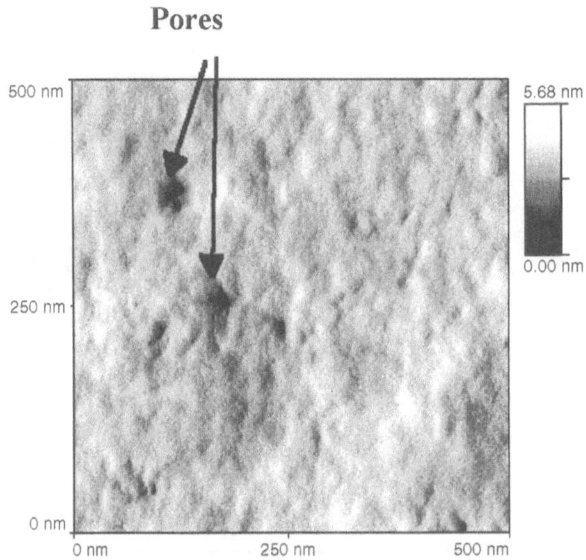

Fig. 2. AFM image of an n-InP sample scanned from 0.0 to 0.785 V (SCE) at $10\,mV \cdot s^{-1}$ in $3\,mol \cdot dm^{-3}$ $(NH_4)_2S$

The surface topography was also examined after a sample had undergone anodization up to 0.7 V. However, no pits were observed in the AFM images taken. In order to investigate whether more prolonged treatment would accentuate pit formation, samples were subjected to repeated cycling between 0.0 V and 0.7 V. Even after such treatment, no obvious pits were observed in the AFM images, although the surface was found to have a much rougher texture than that of an untreated electrode. Thus, the onset of pit formation in the InP surface appears to correspond to the rapid rise in current in the potential region indicated as B in Fig. 1b.

The topography of samples subjected to a potential cycle between 0.0 V and E_U values of 0.82 V (*i.e.* peak potential) and 0.88 V, respectively, was examined, and it is clear from a comparison of the AFM images obtained that the density of pits continues to increase over the potential range from 0.75 to 0.88 V. Thus, it appears that at potentials between 0.5 and 0.75 V etching of the InP occurs with the development of a characteristic surface texture but with no pit formation. Above 0.75 V, pits are formed in the InP surface with a consequent accelerated increase in current. Comparison of AFM images corresponding to potential cycling to 0.82 and 0.88 V shows a change in texture, probably due to the growth of a thicker surface film (35 nm at 0.88 V). The decrease in current following the peak at E_P and the subsequent reduced current on the reverse scan is indicative of the growth of a film. However, the amount of decrease in current is relatively small, consistent with the porosity in the film. The results indicate that films grown under such conditions, while exhibiting some degree of passivity, are not effective as passivating layers.

Film growth at higher potentials

Figure 3 shows a cyclic voltammogram at a scan rate of $10\,mV \cdot s^{-1}$ from 0 to 2.2 and back to 0 V. It exhibits two noteworthy features. Firstly, it is apparent that

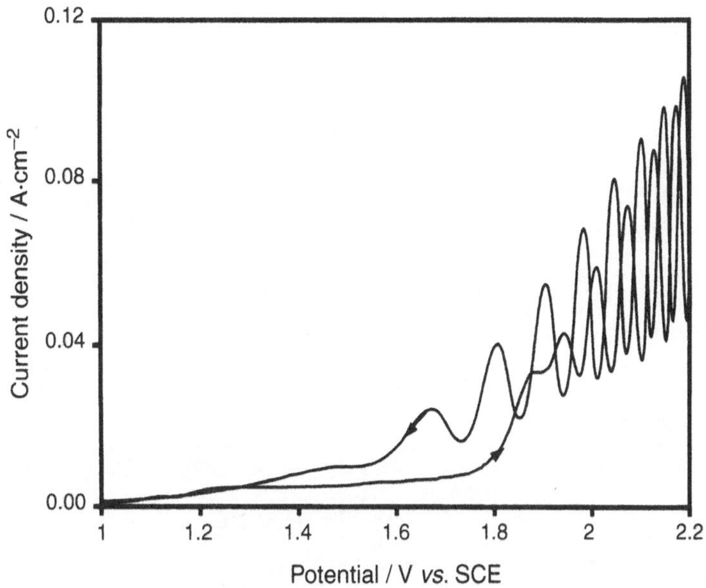

Fig. 3. Cyclic voltammogram of InP between 0.0 and 2.2 V (SCE) in 3 mol · dm^{-3} $(NH_4)_2S$ at a scan rate of 10 mV · s^{-1}

current oscillations occur in the anodic scan at potentials above 1.7 V; this feature will be discussed later. Secondly, the current on the cathodic sweep is, in general, greater than the corresponding (average) current on the anodic scan.

This contrasts with observations at lower potentials (cf. Fig. 1b) and suggests that the films formed at higher potentials are not passivating in nature. Electrodes exposed to potential cycles such as in Fig. 3 and subsequently removed from the cell and examined by scanning electron and optical microscopy clearly showed a pattern of cracking of the surface film. Figure 4 presents a scanning electron micrograph (SEM) surface view of a typical film formed as a result of a cyclic potential sweep between 0.0 and 1.95 V. Extensive cracking of the film is obvious from this micrograph.

To investigate further the origin of this cracking, an InP electrode was subjected to a cyclic potential sweep between 0 and 2.4 V. The electrode was rinsed in deionized water and quickly transferred while still wet to a container in which the relative humidity was maintained at approximately 100%. The sample was then quickly transferred from the container to the optical microscope stage and immediately examined. A time sequence of images obtained as the sample was allowed to dry in ambient laboratory air is shown in Fig. 5. The first image obtained (Fig. 5a) shows that initially the surface was essentially featureless at this magnification, with no evidence of surface cracking. However, after a few minutes (Fig. 5b), cracks appeared in the film. The progression of film cracking was monitored over a period of approximately 20 minutes, and it is obvious from the images in Figs. 5a through 5d taken over this time period that crack formation and broadening continued as the sample was allowed to dry. We attribute the formation of the cracks observed to a shrinkage of the film as it dries. As shown below, the film involved is highly porous, and this is expected to enhance the shrinkage when water is lost by evaporation. Clearly, the surface cracking occurs *ex situ* and is

Fig. 4. SEM showing the characteristic cracking of the film formed after a cyclic potential sweep between 0.0 and 1.95 V (SCE) in 3 mol · dm^{-3} $(NH_4)_2S$ at a scan rate of 10 mV · s^{-1}

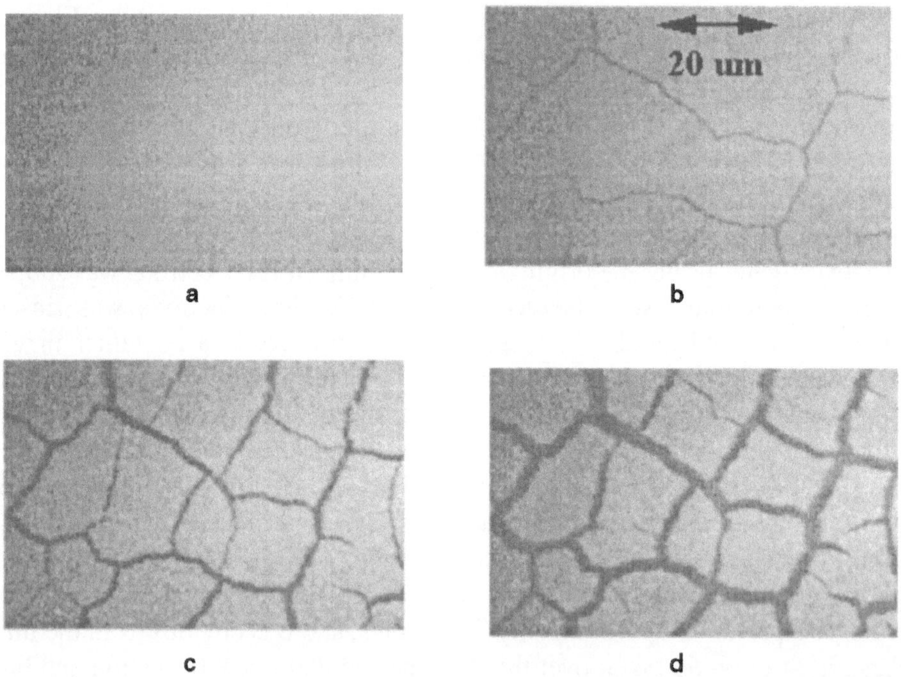

Fig. 5. Time-lapse sequence of optical micrographs showing the evolution of surface cracks on a film as it dries; (a) through (d) were taken sequentially over a 20 min period; the film was formed by cycling the electrode between 0.0 and 2.4 V (SCE) in 3 mol · dm^{-3} $(NH_4)_2S$ at a scan rate of 10 mV · s^{-1}

not a property of the film during the electrochemical treatment. It appears that the cracking occurs due to stress induced by drying of the highly porous film and does not necessarily imply stress in the wet film as grown. We have also observed similar cracking of films formed on p-InP at high anodic potentials. Such cracks have been reported previously [17] in the case of anodization of a p-InP surface in aqueous $(NH_4)_2S$, but their formation was ascribed to stress that occurs in the surface film during anodization. In view of the present results, however, this would not appear to be the origin of this cracking.

Cross sections of the electrode were prepared by cleaving the InP substrate along the (110) crystal plane and allowing the film to break along the cleavage plane of the underlying substrate. The cracks apparent in the surface micrographs are also evident in these cross sections and are observed to extend, typically, through the full thickness of the film.

The elemental composition of films was investigated using both energy dispersive X-ray analysis (EDX) and X-ray photoelectron spectroscopy (XPS). In order to eliminate any effect due to signals from In and P in the underlying InP, samples of the film were detached from the substrate and examined. Spectra obtained from such detached films show that S and In are the predominant elements in the film, whereas P appears to be present only in small quantities. The In:P atomic ratio was calculated for two samples in which the film had been detached. When a correction was made for P assumed to be present as InP, an average value of 0.65 was obtained for the In:S ratio in the film. This is in reasonable agreement with the calculated value of 0.67 corresponding to In_2S_3. XPS measurements were carried out on a sample which had been subjected to a potential sweep from 0.0 to 1.8 V. The In 3d5/2 peak obtained from the surface film is shifted to a higher binding energy relative to the corresponding peak obtained from the substrate. This shift in binding energy was measured to be 0.4 eV and indicates a change in the bonding structure of In: it is consistent with In present as In_2S_3 [31]. In summary, the EDX and XPS measurements suggest that, in agreement with previous reports [17], the film consists predominantly of In_2S_3.

Film porosity

Film thicknesses were obtained from cross-sectional optical and electron micrographs of dry films. The samples were subjected to a cyclic potential scan from 0 V to a series of values of upper potentials greater than 1.7 V (and back to 0 V). The total charge density (Q) passed during the cyclic potential sweep was determined by estimating the integral of the current with respect to time from the total area (forward and backward sweeps) under the cyclic voltammogram curve. This charge was estimated for each of the samples used in the thickness measurement study. Estimates obtained in this way clearly show that the measured film thickness is proportional to the charge passed, and a value of $1.64\,\mu m \cdot C^{-1} \cdot cm^2$ is obtained for the ratio of measured film thickness to charge. The fact that the film thickness increases linearly with the charge passed indicates that a constant percentage porosity is maintained.

A theoretical value for the ratio of film thickness to charge passed during film growth was also estimated. We assume an electrochemical process such as given in

Eq. (1) leading to the formation of an In_2S_3 film and dissolved oxo anions of phosphorous.

$$InP + 3/2\,S^{2-} + 4H_2O + 8h^+ \longrightarrow 1/2In_2S_3 + PO_4^{3-} + 8H^+ \quad (1)$$

The formation of PO_4^{3-} (a P(V) oxo anion) as written in Eq. (1) corresponds to an 8-electron process, whereas, for example, the formation of a P(III) oxo anion such as HPO_3^{2-} corresponds to a 6-electron process. For a compact film, using *Faraday*'s law, the ratio of the thickness d to the charge Q is given by Eq. (2) where $V_{M(In_2S_3)}$ is the molar volume of In_2S_3, n is the number of holes per formula unit of InP, and F is the Faraday constant.

$$\frac{d}{Q} = \frac{V_{M(In_2S_3)}}{2nF} \quad (2)$$

Using a value of $73.17\,cm^3 \cdot mol^{-1}$ for $V_{M(In_2S_3)}$ [32] and assuming $n=8$, a theoretical value of $0.474\,\mu m \cdot C^{-1} \cdot cm^2$ is estimated from Eq. (2) for d/Q, the ratio of (compact) film thickness to charge. As indicated above, an estimate of $1.64\,\mu m \cdot C^{-1}$ was obtained for the corresponding ratio d_e/Q of experimental film thickness d_e to charge, and thus the as-measured film thickness is found to exceed the estimated value for a compact film by a factor of $r=3.46$, indicating that the film is highly porous. Using the estimated value of $r=3.46$, we obtain an estimate of film porosity of approximately 71%. It is clear from Fig. 4 that lateral shrinkage (and hence cracking) of the film occurs. From surface-view micrographs, quantitative estimates of the degree of shrinkage were achieved [33]. Talking this into account we obtain a corrected value of porosity $p=78\%$. As stated above, this assumes an 8-electron process: assuming a 6-electron process leads to a value of $r=2.59$ and a porosity of 71%. Thus, we estimate the porosity to be in the range 71–78%.

It is clear that films formed at potentials above 1.7 V are quite porous. Film growth mechanisms on semiconductors including Si [34], GaAs [35], and InP [36] are often associated with the drift of ionic species through the anodic film under the influence of an electric field. While such a mechanism may operate for film growth at lower potentials in the present InP/S system, at higher potentials we believe that the mechanism of growth involves diffusion of ions through an electrolyte-saturated porous layer rather than ionic transport through a compact film. The porosity of the films formed in this study explains why the surface films do not inhibit current in this region. Thus, for example, currents on the cathodic sweep are larger than the corresponding currents on the anodic sweep. This contrasts with the situation at lower potentials where more compact films are formed, which impose some inhibition to the current flow.

Oscillatory behaviour

As noted earlier, an interesting feature of the anodic potential sweep in Fig. 3 is the presence of current oscillations. Such oscillations appear on the forward sweep of the cyclic voltammograms when the upper limit, E_U, is greater than 1.7 V. They also appear on the reverse sweep when $1.7\,V < E_U < 2.4\,V$. As can be seen from Fig. 3, the current increases rapidly when the potential rises above 1.7 V, and so

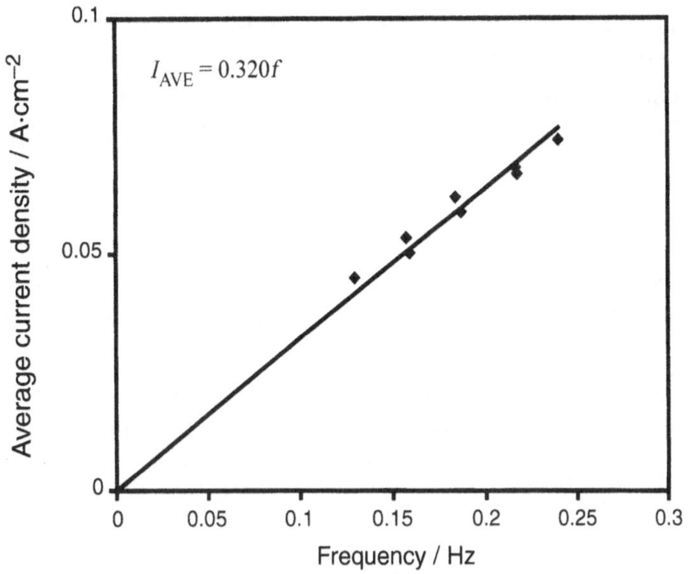

Fig. 6. Average current density plotted against frequency of oscillation under potential sweep conditions; the data were obtained from the cyclic voltammogram shown in Fig. 3

does the frequency of the oscillations. The relationship of the frequency of oscillation to the magnitude of the current was investigated by plotting the average current density against the observed oscillation frequency for a number of voltammograms. The data from one such voltammogram are plotted in Fig. 6 and yield a straight line through the origin.

This indicates that the charge per cycle is constant, with a value of $0.320\,C \cdot cm^{-2}$ given by the slope of the line. Thus, as the potential increases in a voltammogram from 1.7 to 2.4 V and the current increases correspondingly, the charge per cycle remains constant, and the frequency of oscillation increases.

Whereas the frequency range of the oscillations obtained from a given voltammogram is small, increasing the scan rate results, as expected, in an increase in the average current density with a corresponding increase in the frequency. A series of experiments was carried out in which the scan rate was varied from 0.001 to $0.1\,V \cdot s^{-1}$. The measured frequencies of the oscillations are listed in Table 1 and span the range from 0.1 to 1.25 Hz, the charge per cycle remaining constant

Table 1. Frequency range and charge per cycle measured from cyclic voltammograms carried out at various scan rates; the average value of charge per cycle is $0.323\,C \cdot cm^{-2}$ with a standard deviation of $0.010\,C \cdot cm^{-2}$

Scanrate/$V \cdot s^{-1}$	Observed frequency range/Hz	Charge per cycle/$C \cdot cm^{-2}$
0.001	0.10–0.17	0.34
0.005	0.16–0.29	0.32
0.01	0.15–0.32	0.32
0.02	0.29–0.59	0.32
0.05	0.54–0.93	0.33
0.1	0.73–1.25	0.31

within experimental error with an average value of $0.323\,C \cdot cm^{-2}$ and a standard deviation of $0.010\,C \cdot cm^{-2}$. Thus it is clear that, over a significant range of current density and frequency, the charge per cycle remains constant. The reasons for this are currently being investigated, and a more detailed analysis will be given in a future publication.

Oscillations of the current were also observed under constant potential conditions. Again a critical potential exists above which oscillations in the current density were observed. This potential is close to the value above which oscillations are observed in the cyclic voltammograms (*i.e.* approximately 1.7 V). A series of experiments was carried out in which the potential was stepped from open circuit to progressively higher potential values in the range from 1.8 to 2.5 V. Typical results are shown in Fig. 7 in which the potential was stepped to 1.8 and 2.1 V, respectively.

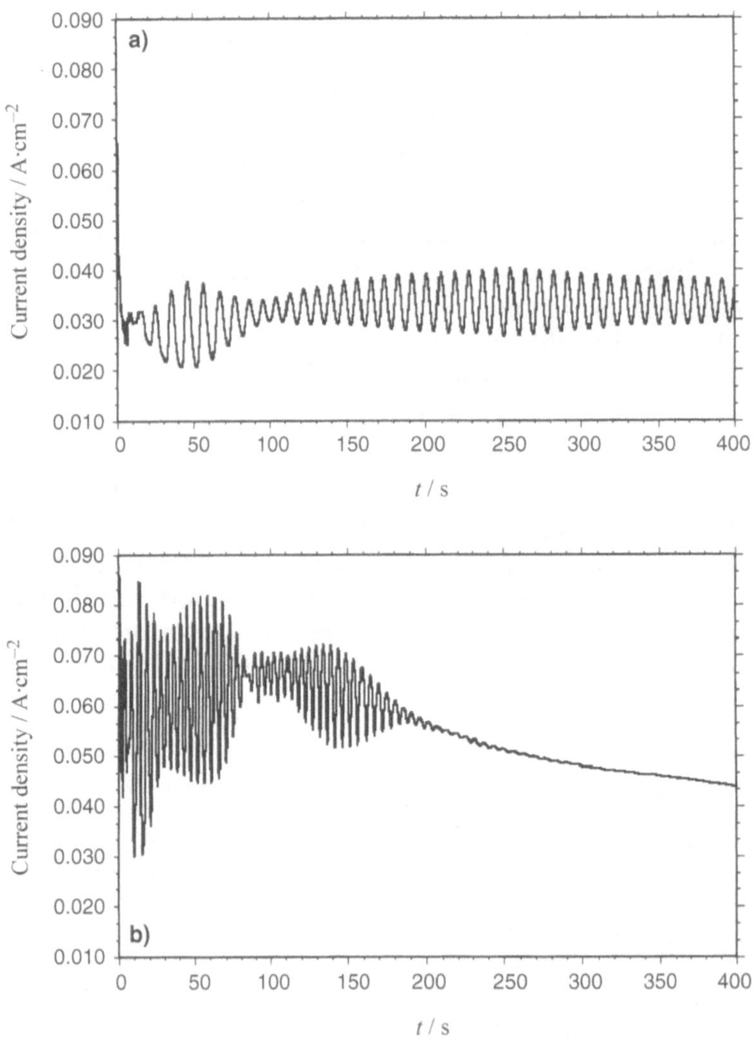

Fig. 7. Variation of current density with time at a constant potential of a) 1.8 and b) 2.1 V (SCE) in $3\,mol \cdot dm^{-3}$ $(NH_4)_2S$

Both curves show typical oscillatory behaviour. It is apparent from Fig. 7 that the average current density is higher after a potential step to 2.1 V and that the duration of the oscillations is longer for the lower imposed potential of 1.8 V. The charge per cycle also remained constant during these constant potential measurements: an average charge per cycle of $0.318\,C \cdot cm^{-2}$ was obtained. This is in good agreement with the value of $0.323\,C \cdot cm^{-2}$ given in Table 1 whose data originate from potentiodynamic measurements. Thus, for all values of the potential where oscillations were observed the average charge per cycle, estimated from the current *vs.* time curves, was found to be approximately constant, independent of the potential, and comparable in value to the constant charge per cycle estimated from the cyclic voltammetric measurements. Oscillations in potential under conditions of constant current were also observed and exhibited a complex signature.

Mechanism of current oscillations

The oscillatory behaviour in the Si/F system is often attributed to growth and dissolution of a thin surface film [37, 38] or changes in the surface roughness [39] of the film grown; *Carstensen et al.* [40] have developed a model which appears to agree well with experimental results. This model is based on an ionic breakthrough mechanism which leads to enhanced localized ion transport to the Si–SiO_2 interface. Under the conditions where oscillations occur in the present study, the InP electrode is covered by a thick (several microns) porous film that continues to thicken with increasing potential or time. This is much different from the Si/F system, in which the anodic films are of the order of tens of nanometres in thickness.

In metal/electrolyte systems, local changes in the *pH* and conditions for the formation and chemical dissolution of a passivating film are often associated with electrochemical oscillations [41, 42]. Some proposed mechanisms involve porous salt films and passivating oxides [43]. Cycling between two interface conditions as a consequence of diffusion or concentration gradients which cause cycling between high and low concentrations of H^+ and other species at the interface has been postulated. The case of copper in acidic chloride solutions is particularly interesting in the context of our results. In that case, *Bassett et al.* [44] have reported a thick porous film (in the order of tens of microns) on the electrode surface during the occurrence of oscillatory behaviour [44]. These authors also report a current-voltage behaviour similar to that observed by us for InP in the present study and postulate that oscillatory behaviour may be a consequence of the formation and dissolution of a thin oxide layer beneath the thick porous film.

In the present study, current densities are relatively high in the region where oscillations are observed. Consequently, relatively large changes in composition within the pores in the vicinity of the InP substrate are expected during current flow. For example, anodic dissolution of InP is accompanied by H^+ formation, and significant changes in *pH* are therefore expected. Similarly, the formation of In^{3+} and oxo anions of phosphorus occurs, and water may even be depleted within the pores of the film adjacent to the InP. Under such conditions, the nucleation and growth of a solid phase resulting in a thin compact film at the interface of the InP and the porous film would not be surprising. This could lead to a rapid reduction in current. After the current had decreased, the interphase region would recover due to the less polarizing influence of the lower current, and eventually the current would again increase. Such a scheme would provide the feedback mechanism essential for oscillatory behaviour.

Thus, although it is quite plausible that the oscillations are due to large changes in composition, including perhaps solid film formation at the interface of the InP and the porous surface film as outlined above, further work is required to explain the very regular and reproducible quantitative behaviour observed. An attractive feature of this proposed mechanism is that it may enable the extension of models developed for Si/F to the present system. We are currently working on a numerical model based on such a mechanism.

Conclusions

The surface properties of InP electrodes were examined following anodization in an $(NH_4)_2S$ electrolyte. An observed current peak in the cyclic voltammogram was attributed to selective etching and film formation, and AFM images revealed surface pitting. Important for the understanding of passivation processes in sulfide solutions, two different types of surface film were observed to form depending on the applied potential. At lower potentials a compact film forms, whereas at higher potentials a transition from compact to porous film formation occurs.

Cracking of the surface film formed at potentials above 1.7 V was observed. The composition of this thick film was identified by EDX and XPS to be In_2S_3. It was demonstrated unambiguously by time-lapse optical microscopy that this cracking is not present when the electrode is removed from the cell but is an artifact of film drying. This is in contrast to the suggestion of *Gao et al.* [14] that similar cracking observed on p-type InP electrodes occurs *in situ* due to stress in the film during the anodization process. It is also clear that the cracking is a direct consequence of the porous, electrolyte-soaked nature of the film. Thus, an easy access pathway appears to exist for diffusion of ions between the substrate and the bulk electrolyte, and film growth is not inhibited. The film thickness was found to increase linearly with the charge passed, and quantitative estimates indicate that a constant percentage porosity of over 70% is maintained throughout the film.

Electrochemical oscillations were observed under three distinct and significantly different conditions: potential sweep, constant potential, and constant current. During potential sweep experiments at various scan rates, the average current density was found to be proportional to the frequency of the oscillations so as to sustain a constant charge per cycle of approximately $0.3\,C\cdot cm^{-2}$. Despite the differences in experimental conditions, current oscillations observed under constant potential conditions also showed a proportionality between average current and frequency and a similar value of approximately $0.3\,C\cdot cm^{-2}$ for the charge per cycle. The detailed mechanism of current oscillations is not clear. It is suggested that it involves large changes in electrolyte composition, including perhaps solid film formation at the interface of the InP and the porous surface film, similar in some respects to oscillatory processes on silicon in fluoride-based electrolytes.

Experimental

The working electrode consisted of (100)-oriented monocrystalline sulfur-doped n-InP with a carrier concentration of approximately $4 \times 10^{18}\,cm^{-3}$. An ohmic contact was established by alloying indium to the InP sample, and the contact was isolated from the electrolyte by means of a suitable varnish. Anodization was carried out in a $3\,mol \cdot dm^{-3}$ aqueous $(NH_4)_2S$ electrolyte, and a conventional three electrode configuration was used for the electrochemical experiments; all potentials are referenced to a saturated calomel reference electrode (SCE).

For cyclic voltammetric measurements and potentiostatic measurements, a CH Instruments Model 650 A electrochemical workstation interfaced to a PC was employed for cell parameter control and data acquisition. For constant current experiments, an EG&G Princeton Applied Research Model 363 potentiostat/galvanostat was used. All electrochemical experiments were carried out at room temperature and in the dark. The surfaces and cross-sections of the anodized samples were examined using a Nikon Nomarski optical microscope, a Joel JSM 840 scanning electron microscope, and a Topometrix atomic force microscope.

Acknowledgements

E. Harvey gratefully acknowledges an Enterprise Ireland Research Scholarship.

References

[1] Shaw DA and Thornton PR (1970) Solid State Electronics **13**: 919
[2] Sandroff CJ, Nottenberg RN, Bischoff JC, Bhat R (1987) Appl Phys Lett **51**: 33
[3] Gerard I, Simon N, Etcheberry A (2001) Appl Surf Sci **175–176**: 734
[4] Schmuki P, Spoule GI, Bardwell JA, Lu ZH, Graham MJ (1996) J Appl Phys **79**: 7303
[5] Eftekhari G (1994) Thin Solid Films **248**: 199
[6] Sumathi RR, Kumar MS, Dharmarasu N, Kumar J (1999) Matls Sci Engineer B **56**: 25
[7] Peng LH, Liao CH, Hsu YC, Jong CS, Huang CN, Ho JK, Chiu CC, Chen CY (2000) Appl Phys Lett **76**: 511
[8] Rotter T, Ferretti R, Mistele D, Fedler F, Klausing H, Stemmer J, Semchinova OK, Aderhold J, Graul J (2001) J Crystal Growth **230**: 602
[9] Elbahnasawy RF, McInerney JG (1999) In: Andricacos PC, Searson PC, Reidsema-Simpson C, Allongue P, Stickney JL, Oleszek GM (eds) Proceedings of Electrochemical Processing in ULSI Fabrication and Semiconductor/Metal Deposition II, PV 99-9. The Electrochemical Society, Proceedings Series, Pennington, NJ, p 242
[10] Yuzer H, Dogan H, Koroglu J, Kocakusak S (2000), Spectrochim Acta Part B **55B**: 991
[11] Bessolov VN, Konenkova EV, Lebedev MV, Zahn DRT (1999) Phys Solid State **41**: 793
[12] Wang Y, Dairici Y, Holloway PH (1992) J Appl Phys **71**: 2746
[13] Pang Z, Song KC, Mascher P, Simmons JG (1999) J Electrochem Soc **146**: 1946
[14] Huh C, Kim SW, Kim HS, Lee IH, Park SJ (2000) J Appl Phys **87**: 4591
[15] Yota J, Burrows VA (1993) J Vac Sci Technol A **11**: 1083
[16] Li ZS, Hou XY, Cai WZ, Wang W, Ding XM, Wang X (1995) J Appl Phys **78**: 2764
[17] Gao LJ, Bardwell JA, Lu ZH, Graham MJ, Norton PR (1995) J Electrochem Soc **142**: L14
[18] Marcu V, Tenne R (1988) J Phys Chem **92**: 7089
[19] Berlouis LEA, Elfick PV, Tarry H (1997) J Chem Soc Faraday Trans **93**: 2291
[20] Berlouis LEA, Peter LM, Greef R, Astles MG (1992) J Crystal Growth **117**: 918
[21] Van Meirhaeghe RL, Cardon F, Gomes WP (1979) Electrochim Acta **24**: 1047
[22] Blackwood DJ, Borazio A, Greef R, Peter LM, Stumper J (1992) Electrochim Acta **37**: 889
[23] Aggour M, Giersig M, Lewerenz HJ (1995) J Electroanal Chem **383**: 67

[24] Langa S, Carstensen J, Tiginyanu, Christophersen M, Foll H (2001) Electrochem Solid-State Lett **4**: G50
[25] Fenollosa R, You H, Chu Y, Parkhutik V (2000) Matls Sci Eng A **288**: 235
[26] Sazou D, Pagitsas M (1992) J Electroanal Chem **323**: 247
[27] Sazou D (1997) Electrochim Acta **42**: 627
[28] Wang C, Chen S, Yu X (1994) Electrochim Acta **39**: 577
[29] Harvey E, Buckley DN (2000) In: Kopf RF, Baca AG, Chu SNG (eds) Proceedings of the 32nd State-of-the-Art Program on Compound Semiconductors PV 2000-1. The Electrochemical Society, Proceedings Series, Pennington, NJ, p 265
[30] Harvey E, O'Dwyer C, Melly T, Buckley DN, Cunnane VJ, Sutton D, Newcomb SB, Chu SNG (2001) In: Chang PC, Chu SNG, Buckley DN (eds) Proceedings of the 35th State-of-the-Art Program on Compound Semiconductors PV 2001-2. The Electrochemical Society, Proceedings Series, Pennington, NJ, p 87
[31] Tao Y, Yelon A, Sacher E, Lu ZH, Graham MJ (1992) Appl Phys Lett **60**: 2669
[32] Lide DR (ed) (2000) CRC Handbook of Chemistry and Physics, 81st edn. CRC Press, NY
[33] Harvey E, Buckley DN, Sutton D, Newcomb SB, Chu SNG, J Electrochem Soc (submitted)
[34] Rappich J (2000) Microelectronics Reliability **40**: 815
[35] Spitzer SM, Schwartz B, Weigle GD (1975) J Electrochem Soc **122**: 397
[36] Robach Y, Joseph J, Bergignat E, Hollinger G (1989) J Electrochem Soc **136**: 2957
[37] Dini D, Cattarin S, Decker F (1998) J Electroanal Chem **446**: 7
[38] Cattarin S, Chazalviel JN, Da Fonseca C, Ozanam F, Peter LM, Schlichthorl G, Stumper J (1998) J Electrochem Soc **145**: 498
[39] Nast O, Rauscher S, Jungblit J, Lewerenz HJ (1998) J Electroanal Chem **422**: 169
[40] Carstensen J, Prange R, Foll H (1999) J Electrochem Soc **146**: 1134
[41] Russell P, Newman J (1986) J Electrochem Soc **133**: 2093
[42] Rush B, Newman J (1995) J Electrochem Soc **142**: 3770
[43] Beck TR (1982) J Electrochem Soc **129**: 2412
[44] Bassett MR, Hudson JL (1990) J Electrochem Soc **137**: 922

Received October 16, 2001. Accepted (revised) December 21, 2001

A Study of $PbTiO_3$ Crystallization in Pure and Composite Nanopowders Prepared by the Sol-Gel Technique

Marian Čerňanský*, **Přemysl Vaněk**, **Karel Král**, and **Radmila Krupková**

Academy of Sciences of the Czech Republic, Institute of Physics, CZ-18221 Prague, Czech Republic

Summary. In this investigation the crystallization of $PbTiO_3$ upon annealing of pure nanopowders and $PbTiO_3$–SiO_2 (1:1 v/v) nanocomposite powders prepared by the sol-gel technique was studied. Using X-ray diffraction phase analysis, the start of $PbTiO_3$ crystallization in pure $PbTiO_3$ powders was detected at 400°C. Distinct crystallization of $PbTiO_3$ in $PbTiO_3$–SiO_2 nanocomposites starts at 700°C, whereas SiO_2 remains amorphous. There are indications that an interface interaction between the $PbTiO_3$ and the SiO_2 phase plays an important role in hindering the crystallization of $PbTiO_3$. The particle size (size of coherently scattering regions) was estimated from the broadening of the X-ray diffraction line profiles. The average size of $PbTiO_3$ nanocrystallites increases with temperature and time of annealing, the influence of temperature being more significant than that of the annealing time. Differential scanning calorimetry confirmed the results of the X-ray diffraction with respect to the start of the crystallization. Laser beam scattering and scanning electron microscopy provided the statistical distribution of the grain size and the morphology of the powder grains, showing that each grain of the powders contains several nanocrystallites (coherently scattering regions).

Keywords. Calorimetry; Crystallization; Nanostructures; Sol-gel; X-Ray diffraction.

Introduction

Ferroelectric materials with perovskite crystal structure have several interesting properties which are used in technical applications. Their piezoelectricity is used as a base for electromechanical transducers employed as ultrasonic sensors and generators; detectors of infrared radiation are also based on this property. Ferroelectrics have high dielectric permittivity which is favourable for the fabrication of capacitors of very small size [1]. The transparency of some ceramic ferroelectrics varies under the influence of an electric field, making them promising devices for electro-optic applications. Another important property of these materials is spontaneous polarization which can be reversed by an external electric field [2]. The ability of ferroelectric materials to switch their polarization direction between two

* Corresponding author. E-mail: cernan@fzu.cz

stable polarized states provides the basis for binary-code based nonvolatile ferroelectric random-access memories [3].

The important point is that the physical and technological properties of the perovskite ferroelectrics depend on the size of the crystallites [4], making the preparation of these materials in the form of nanocrystals and nanocomposites of interest. Generally, nanocrystalline materials have large surface-to-volume ratios, and their various properties such as stability, melting temperature, sintering ability, electronic structure, *etc.* depend on the size of the nanocrystallites [5]. Moreover, the second component (matrix, heterogeneous admixture) in nanocomposites can influence significantly the properties of the first (active) component (preferred orientation [6], stability of phases [7], and application properties, *e.g.* ionic conductivity [8]).

The purpose of this paper is the study of the $PbTiO_3$ crystallization in pure nanopowders and $PbTiO_3$–SiO_2 (1:1 v/v) nanocomposite powders prepared by the sol-gel technique. The evolution of nanocrystallite particle size at crystallization was estimated from the X-ray diffraction line profile analysis.

Results and Discussion

Samples of $PbTiO_3$ nanopowders and $PbTiO_3$–SiO_2 (1:1 v/v) nanocomposites were annealed at various temperatures for various times. The X-ray diffraction patterns in Figs. 1 and 2 indicate that the crystallization starts in pure $PbTiO_3$ powders at 400°C and in $PbTiO_3$–SiO_2 nanocomposites at 700°C. The SiO_2 phase remains amorphous even after annealing at 800°C for one hour. Differential scanning calorimetry (DSC) agrees with X-ray diffraction concerning the start of crystallization – the first detectable DSC anomaly at about 490°C indicating the existence of a cubic-to-tetragonal phase transition in $PbTiO_3$ (*i.e.* the presence of a crystalline phase) was found in pure $PbTiO_3$ samples annealed at 500°C and in $PbTiO_3$–SiO_2 samples annealed at 800°C. This is comparable with the crystallization

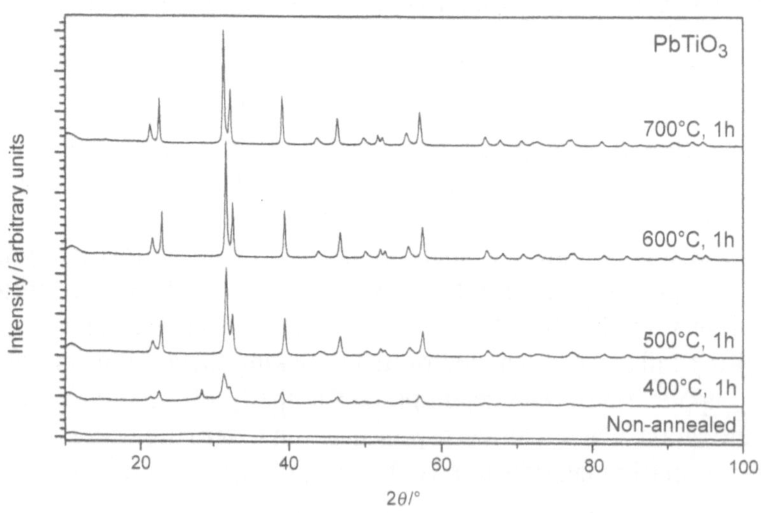

Fig. 1. Powder diffraction patterns of annealed pure $PbTiO_3$ powders

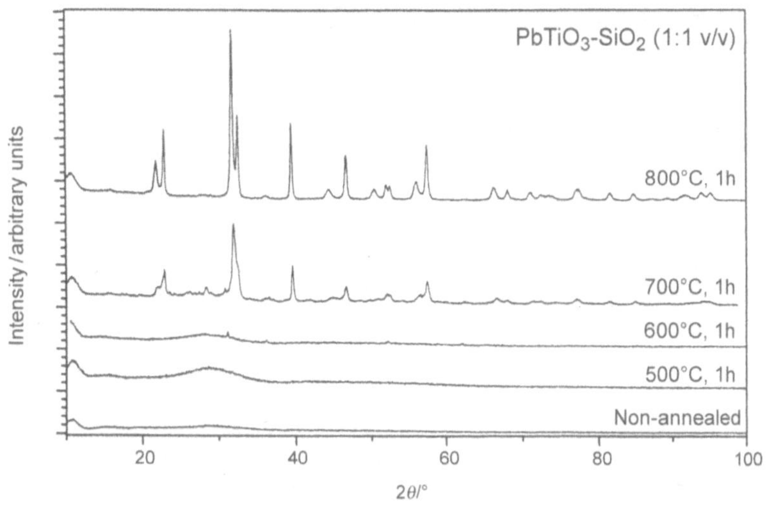

Fig. 2. Powder diffraction patterns of annealed $PbTiO_3$-SiO_2 nanocomposites

temperature (800°C) of $PbTiO_3$ in $PbTiO_3$–SiO_2 glass ceramics with a high content of SiO_2 ($PbTiO_3$:SiO_2 = 8.4:91.6 mol%) prepared by the sol-gel technique [9]. In contrast to amorphous SiO_2 in our nanocomposites, SiO_2 crystallized as cristobalite at 800°C in the glass ceramics mentioned above [9]. The hindering effect of a heterogeneous admixture on $PbTiO_3$ crystallization was also observed in $PbTiO_3$–SiO_2–B_2O_3 glass ceramics prepared by the sol-gel process [10, 11]. The crystallization temperature of $PbTiO_3$ increased with rising content of glassy phase [10, 11]. A similar effect was observed in $PbTiO_3$–Al_2O_3 nanocomposite thin films and, to much less extent, in $PbTiO_3$–Al_2O_3 multilayers [12] where the interface area is much smaller than in nanocomposites. The above data imply that the interface interaction between the components of nanocomposites plays an important role in influencing the crystallization of $PbTiO_3$ (and also SiO_2) phases. This is in agreement with the fact that interface interaction can stabilize an amorphous phase at the interface [7, 8].

In Tables 1 and 2, the parameters of annealing are shown together with the corresponding values of crystallite size *D* (size of coherently scattering regions) and values of microstrain *e* for $PbTiO_3$ powders and $PbTiO_3$–SiO_2 nanocomposites. The values of crystallite size taken from Tables 1 and 2 are plotted in Fig. 3. From Tables 1 and 2 and Fig. 3 it can be concluded that the size of nanocrystallites increases at higher temperature and time of annealing, the influence of temperature being more significant than that of the time. This is also illustrated in Fig. 4 for both nanopowder and nanocomposites. It should be stressed that X-ray diffraction line profile analysis estimates the size of coherently scattering regions, *i.e.* the size of (nano)crystallites. Each grain of powder can contain several nanocrystallites.

The values of the microstrain *e* (Tables 1 and 2) are relatively small compared to the values obtained for other nanostructured materials and other methods of preparation. For example, $e = 0.013$ was found for CuO prepared by ball milling [14], values of 0.002 and 0.010 were determined for nanocrystalline Pd obtained by the gas condensation method [15]. The sol-gel method results in smaller

Table 1. Crystallite size D and microstrain e in $PbTiO_3$

Annealing temperature $t/°C$	τ/h	D/nm	$e \times 10^4$
400	0.5	26	12.7
400	1	28	10.1
500	0.5	32	5.0
500	1	35	14.9
500	2	39	2.5
500	4	43	4.2
600	0.5	72	3.9
600	1	81	5.1
700	0.5	82	4.6
700	1	108	8.6

Table 2. Crystallite size D and microstrain e in $PbTiO_3$–SiO_2

Annealing temperature $t/°C$	τ/h	D/nm	$e \times 10^4$
700	0.5	36	9.5
700	1	38	11.3
700	2	40	13.8
700	4	44	11.9
800	1	42	10.8

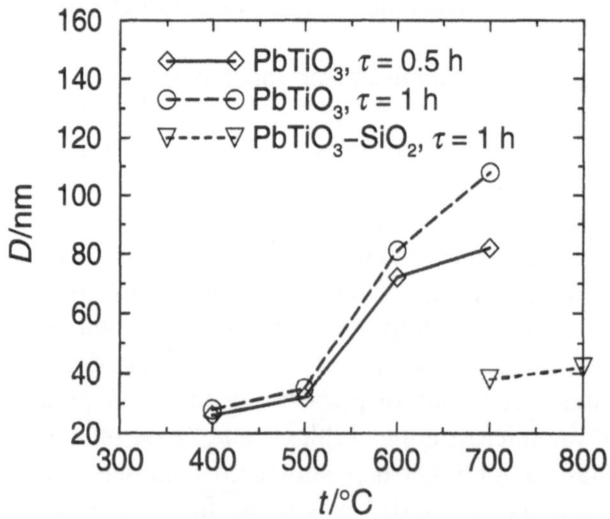

Fig. 3. Dependence of particle size (coherently scattering region) D on the temperature t of annealing at different annealing times τ

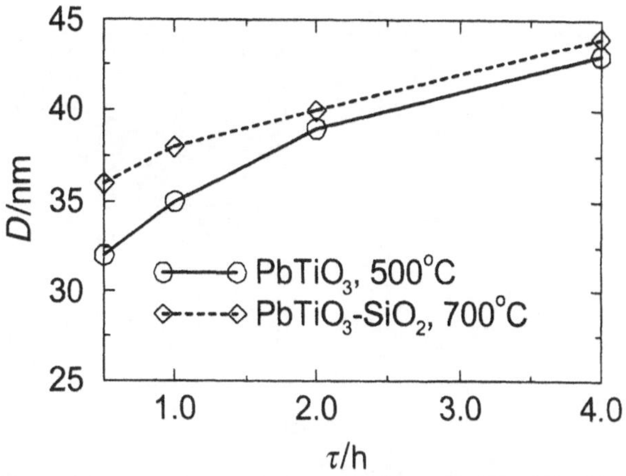

Fig. 4. Dependence of particle size D on annealing time τ

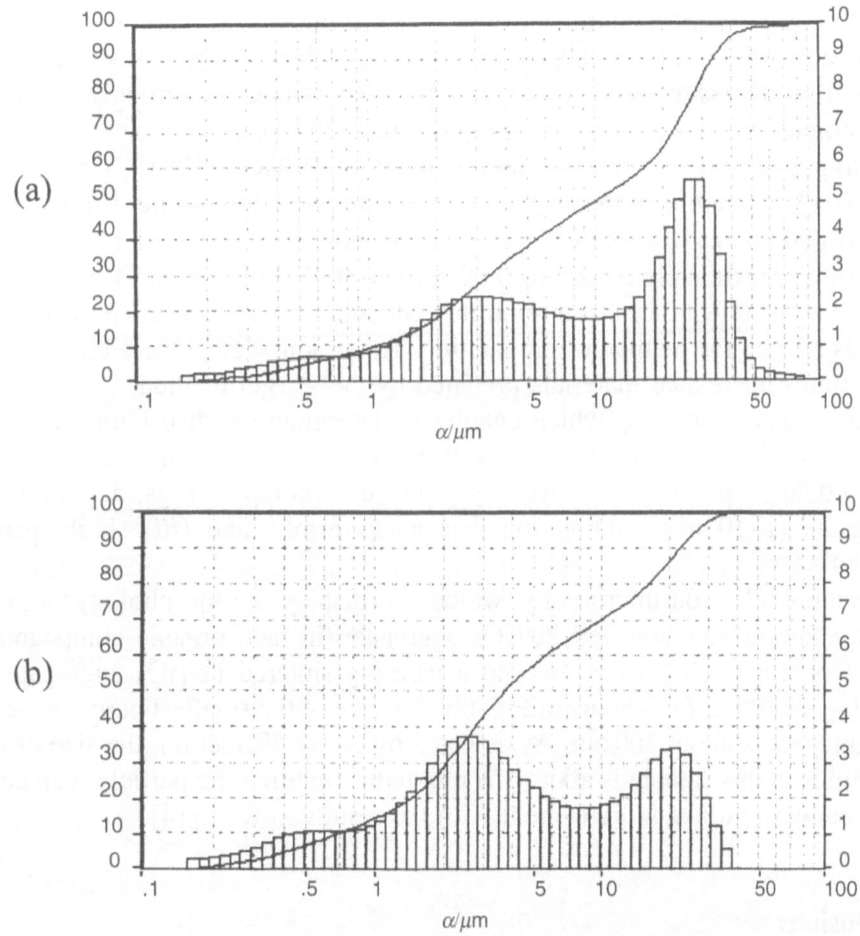

Fig. 5. Results of laser beam scattering experiments; the grain size distribution (relative counts of particles within a given size range) is presented in the form of a histogram (right vertical scale), the integral distribution curve (left vertical scale) is given as a solid line; α: linear grain size; (a) low ultrasound intensity, no detergent used; (b) high ultrasound intensity, detergent used

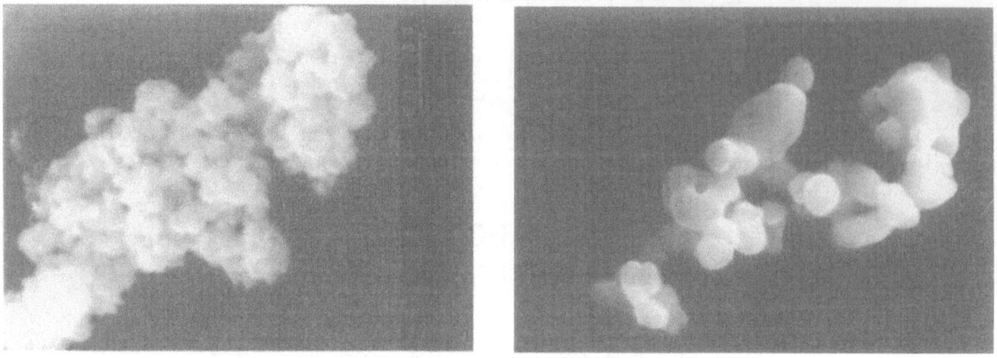

a b

Fig. 6. Scanning electron micrograph of $PbTiO_3$ powder (a) dried for 1 h at 200°C and (b) annealed for 3 h at 800°C; the width of the photograph corresponds to a distance of 4.5 μm

microstrains in nanocrystalline powders than the above mentioned methods. Comparable values of $e=0.0003$–0.0015 were obtained in nanocrystalline Cu prepared by the electrodeposition technique [16]. Annealing strongly reduces the microstrains in the samples. In our procedure, the ratio of the temperature of annealing (400–800°C) to the melting point of $PbTiO_3$ (1281°C) ranges from 0.43 to 0.69 (for temperatures in K). A significant reduction of microstrains in electrodeposited nanocrystalline Cu was observed for a ratio of 0.31 [16]. However, this ratio approximately equals to 0.69 for cryomilled nanocrystals of Zn [17]. It seems that the effect of annealing on the reduction of microstrains in nanocrystals depends also on the method of preparation. To our knowledge there are no data on microstrains for related materials prepared by the sol-gel method.

Laser beam scattering, which enables to determine the distribution of powder grain sizes (Fig. 5), shows three types of grain sizes: (*i*) agglomerates (10–50 μm) which could be partially destroyed by ultrasound (compare Figs. 5a and 5b), (*ii*) aggregates (1–10 μm) holding together much firmly, and (*iii*) single particles ($<1\ \mu m$).

Scanning electron microscopy, which can display the morphology of powder grains, shows a non-annealed $PbTiO_3$ aggregate (in fact, organic compounds are present in it as well) in Fig. 6a and a partially sintered $PbTiO_3$ aggregate after annealing at 800°C for 3 hours in Fig. 6b. The size of the smallest particles forming the aggregate is about 300 nm. As detected by X-ray diffraction, the size of nanocrystallites in this sample is about 159 nm, indicating that the particles can contain a few nanocrystallites in agreement with the results in Ref. [14].

Conclusions

The following conclusions can be drawn from the obtained results:

(*i*) The rate of $PbTiO_3$ crystallization from amorphous powders prepared by the sol-gel technique is stimulated by increased annealing temperatures and to a lower extent, by prolonged annealing time.

(*ii*) The presence of SiO_2 in $PbTiO_3$–SiO_2 (1:1 v/v) composite powders increases the minimum temperature necessary for crystallization by about 300°C in comparison with pure $PbTiO_3$.

(*iii*) Interface interaction between grains of $PbTiO_3$ nanocrystals and an amorphous SiO_2 matrix obviously plays a role in this phenomenon. In order to determine the character of the interaction, further studies have to be performed.

Experimental

$PbTiO_3$ nanopowders and $PbTiO_3$–SiO_2 (1:1 v/v) nanocomposite powders were prepared by the standard sol-gel method [13]. The sol was synthesized by dissolution of water-free lead acetate (trihydrate dried in vacuum at 70°C for 5 h), titanium tetrabutoxide, and silicon tetraethoxide (in the case of the composite) in *n*-butanol. The reaction mixture was stirred and refluxed for 2 h. All operations were performed under dry N_2. The obtained sol was clear and without any solid residues. After cooling to room temperature the sol was rapidly hydrolyzed by a mixture of water and *n*-butanol (1:5 v/v) The resulting white suspension was dried in air at room temperature and then at 140°C (or additionally at 200°C). The obtained white powder was further pulverized in an agate mortar and pressed into a quartz cell for X-ray examination.

X-Ray diffraction measurements were performed with a powder diffractometer using CuK_α radiation from the X-ray source equipped with a rotating anode. The diffractometer was operated with a focusing crystal monochromator in the diffracted beam path and a NaI(Tl) scintillation detector. The fitting procedure of measured diffraction profiles indicated that these profiles could be well approximated by analytical functions of the *Cauchy* type. The same result was obtained for diffraction profiles of the standard sample (tungsten powder). This standard was used to eliminate the instrumental broadening. The *Cauchy* type of measured and standard profiles implies the simple formula $B - b = \beta$ where B, b, and β are the half-widths of the measured profile, the instrumental profile, and the intrinsic (pure) diffraction profile, respectively. An analog linear relation can be assumed for the physical components of the intrinsic diffraction profile. Consequently, the *Williamson-Hall* plot [18] in the linear form $(\beta \cdot \cos\theta)/(K \cdot \lambda) = (1/D) + (4 \cdot e \cdot \sin\theta)/(K \cdot \lambda)$ was used where the half-width of diffraction profile β is given in radians, λ is the wavelength of the X-ray radiation, θ is the *Bragg* angle, D is the crystallite size, $e = \Delta d/d$) is the microstrain (d: interplanar spacing), and K is the *Scherrer* constant depending mainly on the shape of the crystalline particles. Its values are close to 1 [18], and this value was used in our calculations.

The DSC measurements were performed using a Perkin-Elmer DSC7 calorimeter at a scanning rate of 10 K/min. Powder samples (typically 30–40 mg) were placed in sealed aluminum pans, and N_2 (25 cm^3/min) was used as purging gas. Scanning electron microscopy was performed with a JEOL JXA 733 electron microprobe. Laser beam scattering (Fritsch Analysette 22) was used to determine the distribution of powder grain sizes. The powder was dispersed in pure H_2O or in H_2O containing detergent and agitated by ultrasound.

Acknowledgements

This work was supported by the grants IAA1010113, RN 1998 2003 014, NSC 89-211-M-005-022, GA CR 202/96/0425, and GA CR 202/00/1245. We are grateful to *Karel Jurek* for SEM observations, to *Pavel Lejček* and *Vladimír Železný* for fruitful discussions, and to Fritsch company for providing measurement time.

References

[1] Le Marrec F, Fahri R, El Marsi M, Dellis J-L, Ariosa D, Karkut MG (2001) Ferroelectrics **254**: 1
[2] Trainer M (2001) Am J Phys **69**: 966

[3] Auciello O, Scott JF, Rames R (1998) Physics Today, July 22
[4] Tanaka M, Makino Y (1998) Ferroelectric Lett **24**: 13
[5] Wang Z, Saxena SK, Pischedda V, Liermann HP, Zha CS (2001) J Phys Condens Matter **13**: 8317
[6] Buršík J, Vaněk P, Kužel R, Studnička V, Železný V (2001) J Eur Cer Soc **21**: 1503
[7] Uvarov NF, Vaněk P (2000) J Mater Synth Process **8**: 319
[8] Uvarov NF, Vaněk P, Yuzyuk YuI, Železný V, Studnička V, Bokhonov BB, Dulepov VE, Petzelt J (1996) Solid State Ionics **90**: 201
[9] Qi L, Ma J, Cheng H, Zhao Z (1996) J Mater Sci Lett **15**: 1074
[10] Zhai JW, Yao X, Zhang LY (2000) J Electroceramics **5**: 211
[11] Zhai JW, Yao X, Zhang LY (2001) J Inorg Mater **16**: 147
[12] Buršík J, Vaněk P, Studnička V, Ostapchuk T, Buixaderas E, Petzelt J, Krupková R, Březina B, Peřina V (2000) Ferroelectrics **241**: 191
[13] Klein LC (1996) Processing of nanostructured sol-gel materials. In: Edelstein AS, Cammarata RC (eds) Nanomaterials: Synthesis, Properties and Applications. Institute of Physics Publishing, p 145
[14] Stewart SJ, Borzi RA, Punte G, Mercader RC, Garcia FJ (2001) J Phys Condens Matter **13**: 1743
[15] Fitzsimmons MR, Eastman JA, Müller-Stach M, Wallner G (1991) Phys Rev **B44**: 2452
[16] Lu L, Tao NR, Wang LB, Ding BZ, Lu K (2001) J Appl Phys **89**: 6408
[17] Zhang Z, Wang H, Narayan J, Koch CC (2001) Acta Mater **49**: 1319
[18] Klug HP, Alexander L (1974) X-Ray Diffraction Procedures for Polycrystalline and Amorphous Materials, 2nd edn. Wiley, New York

Received October 4, 2001. Accepted (revised) December 14, 2001

Self-Organization of Magnetic Nanoparticles and Inclusion of Hydrogen by Borohydride Reduction

Iovka Dragieva[1,*], **Christina Deleva**[1], **Mladen Mladenov**[1], and **Ivania Markova-Deneva**[2]

[1] Central Laboratory of Electrochemical Power Sources, Bulgarian Academy of Sciences, BG-1113 Sofia, Bulgaria
[2] Chemical Technology and Metallurgical University, BG-1756 Sofia, Bulgaria

Summary. Different structures of the interglobular space or voids between self-organized nanoparticles lead to differences in the measurable magnetic properties of single-domain particle chains of similar composition, grain size, and amorphous structure of the single globules. The volumes and radii of nanoparticles obtained by application of a magnetic field (3 to 15 nm) are larger than those determined without application of a magnetic field during the borohydride reduction process. Two types of hydrogen containing nanotubes with diameters of up to 2 (small-size containers) and 5 nm (large-size containers) are produced using as a driving force the domain wall formation energy between ferromagnetic nanoparticles with quantum size effected dimensions prepared by this reduction method at room temperature and ambient atmosphere. Nanoscale hydrogen containers can be used instead of *Me*H nanoparticle electrodes as perfect energy charge transfer media of high efficiency (close to 100%) using Li ion electrolytes. No influence on the electrode temperature and no participation of OH^- and H_2O in the main charge/discharge transfer reactions were observed.

Keywords. Self-organization; Ferromagnetic nanoparticles; Borohydride reductions; Charge transfer; Thermodynamics.

Introduction

The aim of this study is to continue our investigations on nanoscaled tubes formed from single-domain ferromagnetic nanoparticles produced by borohydride reduction, containing inside hydrogen clusters arranged as tubes or channels of different diameters (2 or 5 nm) [1, 2]. The possibility for application of these nanoscale tubes not as hydrogen storage elements or *Me*H electrode materials, but with regard to their application in Li ionic solid state batteries without participation of OH^- groups or H_2O in the main charge/discharge transfer reactions is discussed.

* Corresponding author. E-mail: iovka@cleps.bas.bg

Results and Discussion

Self-organization of nanoparticles in chains

The borohydride reduction process leads to the formation of single-domain (amorphous or crystalline [3]) nanoparticles whose shape, size, grain size distribution, structure, and magnetic and conductive properties depend on technological parameters such as metal salts used, complex-forming agents, time of mixing and hydrodynamic regime, *pH* value of the solutions used, application of a DC magnetic field during the reduction process, *etc*. The use of a magnetic field during the preparation of ferromagnetic amorphous nanoparticles by borohydride reduction produces no change in the kind of particles formed in terms of structure or shape, but affects the configuration of the interparticle space or voids of the chains obtained as a result of nanoparticles self-organization. Curling and buckling nets or chains built from single domain globules trap the hydrogen delivered during the borohydride reduction process into nano-dimensional cages [4]. The experimental data obtained from BET absorption isotherms show that depending on whether a DC magnetic field is applied or not, interglobular spaces of definite dimensions and structures appear [4]. Thus, irrespective of the presence of the closest packing of nanoparticles along the chains, the applied magnetic field decreases the void volume (the inside diameter of the channels filled with hydrogen is 2 nm), and as a result, the chain becomes denser. As established earlier [5–11], the different structures of the interglobular spaces lead to differences in the measurable properties of single-domain particle chains of similar composition, grain sizes, and magnetic amorphous structures of the single globules.

The basic element proposed for nanoparticle chains obtained in a DC magnetic field is a *pentahedron*, whose sixth globule (with flexible volume) is situated above the pentahedral level [5, 6, 8] and which exhibits a *tetrahedral* interparticle space (cross-sectional shape of hydrogen channels).

The basic element proposed for nanoparticle chains obtained in absence of a magnetic field is a *hexagonal* ring with the seventh globule incorporated in its center (on the same level with the remaining six surrounding globules [5]) and having an *octahedral* interglobular space.

Particles with radii smaller than 10 nm

As the particle size decreases towards some critical particle diameter, the formation of domain walls becomes energetically unfavourable, and the resulting species are called single-domain particles. Changes in the magnetization can no longer occur through domain wall motion and require the coherent rotation of spins instead. The activation volume of the cluster or of the nanoparticle is an estimate of the volume over which spins act coherently. Coherent rotation of all spins is assumed when the activation volume is smaller than the physical volume of the switching unit due to incoherent reversal mechanisms. One of the most interesting aspects of the behaviour of a nanometer size magnetic particle is the fact that the magnetic north and south poles may suddenly interchange due to quantum tunneling. In this case, the magnetic moment of the particles behaves as a quantum object rather than a

Table 1. Influence of DC magnetic field application during the synthesis of nanoparticles by borohydride reduction on grain size and content of hydrogen and boron [12]

Magnetic field	$(FeCo)_xB_yH_z$		$(SmCo_5)_xB_yH_z$		$(NiCo)_xB_yH_z$		$(Fe_2Si)_xB_yH_z$	
	no	yes	no	yes	no	yes	no	yes
$SSA/m^2 \cdot g^{-1}$	87	59	88	75	11	41	24	30
Particle radius/nm	5.7	8.5	7.2	8.6	47	12	26	21
Hydrogen content/at.%	0.27[a]	0.44[a]	27	29	18.1	12.5	11.2	16.9
Boron content/at.%	4.3[a]	3.6[a]	25.2	29.4	23.6	25.8	14.2	10.7
Number of atoms	–	–	$2 \cdot 10^5$	$4 \cdot 10^5$	$5 \cdot 10^7$	$1 \cdot 10^6$	$1 \cdot 10^7$	$6 \cdot 10^6$

[a] Data from Ref. [9]

classical vector. The two equivalent but opposite orientations of the magnetic moment are separated by an energy barrier which varies linearly with the volume of the particle. The cluster's magnetic moment is decoupled from the lattice and fluctuates thermally in direction, all orientations being equally favourable. When placed in a magnetic field, a nonzero-averaged magnetic moment develops and aligns with the applied external field [12].

Consequently, the measured volumes and radii of nanoparticles obtained under application of a DC magnetic field (Table 1) are larger than those of nanoparticles obtained in the absence of a magnetic field. In such spherical nanoparticles (*e.g.* from 3 to 15 nm; Table 1, $(FeCo)_xB_yH_z$ and $(SmCo_5)_xB_yH_z$), almost 50% of the atoms are on the surface of the particle, and coherent spins can aligns with the external magnetic field, thus increasing the measured volume of the single-domain particles.

Data published earlier have shown that the penetration of gaseous phases like ammonia in the case of gaseous nitrogenation of iron–boron nanoparticles through hydrogen channels with tetrahedral structure is slower and time limited at 400°C in comparison to the same process through hydrogen channels with octahedral structure or with a twice as large cross section (5 nm) [5]. It has also been proved that the smaller nanovoids (with diameters around 2 nm and filled with higher content of hydrogen) are elastic and cannot be subjected to 100 MPa press loading in contrast to the larger voids (5 nm, containing half the amount of hydrogen) [5, 9, 10, 11].

The different magnetic properties – higher coercive force, higher squareness, and lower magnetization (or the opposite) – for both kinds of chains were estimated from specific self-organization phenomena and differences in volume packing density (0.96 and 0.74, respectively, along the chains) [7, 9, 12]. Both types of chains can be included in magnetic lack coatings as longitudinal inside perpendicular to the surface (LIPS) oriented metal magnetic particle layers for high density hard disks production [12].

Thermodynamic analysis

Summarizing our experience on $Me_xB_yH_z$ nanoparticle preparation by borohydride reduction and their application as *Me*H electrodes [13, 14] tested with 6 *M* KOH electrolyte, we decided to avoid the charge transfer between nanoparticle surfaces and electrolyte. The fundamental thermodynamic relationships determining the cell voltage and the basic types of electrode reactions in the case of hydrogen or charged hydrogen groups are as follows:

$$H + e^- \rightarrow H^- \tag{1}$$

$$H^- - e^- \rightarrow H \tag{2}$$

$$H - e^- \rightarrow H^+ \tag{3}$$

$$H^+ + e^- \rightarrow H \quad (\textit{Volmer}) \tag{4}$$

$$H + H \rightarrow H_2 \quad (\textit{Tafel}) \tag{5}$$

$$H^+ + e^- + H \rightarrow H_2 \quad (\textit{Heyrovsky}) \tag{6}$$

The driving force for all reactions is the difference between the standard *Gibbs* free energies of the products and those of the reactants [2]. The thus estimated data *vs.* temperature for reactions according to Eqs. (1) to (6) are shown in Fig. 1.

The probabilities of Eqs. (1), (3), (5), and (6) at room temperature and the required catalysts, temperature, or potentials for Eqs. (2) and (4) can be well estimated. To use the reversibility of the above reactions with a participation of hydrogen or charged hydrogen groups and to prepare *Me*H electrodes and produce batteries on that basis is a difficult technological problem with a high degree of complexity, especially in terms of water elimination.

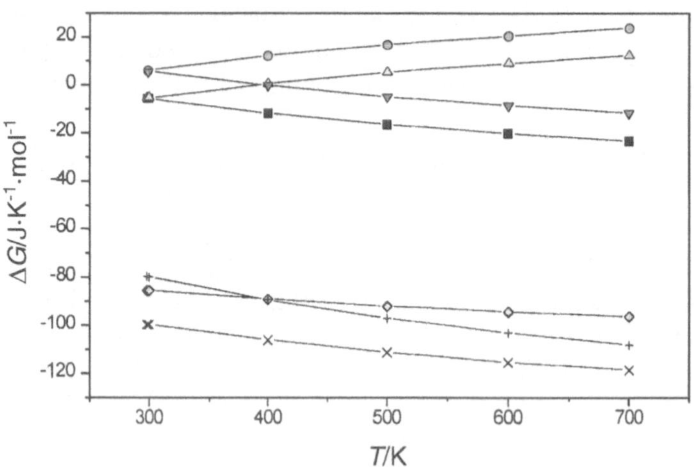

Fig. 1. Standard *Gibbs* free energies *vs.* temperature; Eq. (1): ■, Eq. (2): ●, Eq. (3): △, Eq. (4): ▼, Eq. (5): ◇, Eq. (6): +, Eq. (9): ×

Conclusions

Our hypothesis for the working model in the case of nanoscale hydrogen container (or ferromagnetic nanotubes for storage of hydrogen) applications in Li batteries is as follows:

(*i*) For the reactions taking place on the surface of the hydrogen storage magnetic nanotubes (or nanoscale containers for hydrogen), the charge carrier is the lithium ion according to Eqs. (9), (10), and (11) (cf. Experimental).
(*ii*) The character of the Li–H bond obtained as a result of the electrochemical reactions of stored hydrogen with $LiClO_4$ electrolyte through the net of single-domain nanoparticles is typically covalent, and charge/discharge potentials influence its formation and polarization.
(*iii*) The bond between lithium and hydrogen is stronger (bond strength: 238.049 kJ/mol) than that between lithium atoms (bond strength: 110.21 kJ/mol) [16], and the dispersion of lithium atoms on the nanotube surface is a real fact; however, they do not form pair junctions as thin films on the surface of nanoscaled containers.
(*iv*) Nanotubes filled with hydrogen can be applied as a perfect charge transfer energy media with very high efficiency (close to 100%) with no need of mass transfer of hydrogen or hydrogen containing atomic groups as has been proven for the case of Li ions as charge carriers.

The mechanism of charge/discharge of magnetic chains and nanotubes filled with hydrogen in non-aqueous electrolytes is presently being investigated by IR spectroscopy; first results are very promising.

Experimental

Nanoscale containers for hydrogen or hydrogen-filled channels inside ferromagnetic chains were prepared by the reduction of aqueous metal salt solutions with an aqueous solution of $NaBH_4$ according to Eqs. (7) and (8) [15].

$$NaBH_4 + 2H_2O \rightarrow NaBO_2 + 4H_2\uparrow \qquad (7)$$

$$BH_4^- + 2M^{2+} + 2H_2O \rightarrow 2M\downarrow + NaBO_2 + 4H^+ + 2H_2\uparrow \qquad (8a)$$

$$BH_4^- + H_2O \rightarrow B\downarrow + OH^- + 2.5H_2\uparrow \qquad (8b)$$

Two types of hydrogen containers (small-size container: diameter up to 2 nm; large-size container: up to 5 nm) were produced using as a driving force the domain wall formation energy between ferromagnetic nanoparticles with quantum size effected dimensions prepared by borohydride reduction. Both types of containers are built up from single domain metallic nanoparticle nets, anisotropically self-organized, forming channels of different diameters at the inside with different content of hydrogen and different specific surface areas.

Experiments with aqueous electrolyte (6 M KOH)

The samples were prepared as electrodes by mixing nanoscaled particles (and hydrogen containers, respectively) with teflonized acetylene black (TAB-2) in a weight ratio of 50:50 as anodic and NiOOH as cathodic materials; Hg/HgO was used as a reference electrode. The experimental electrodes were

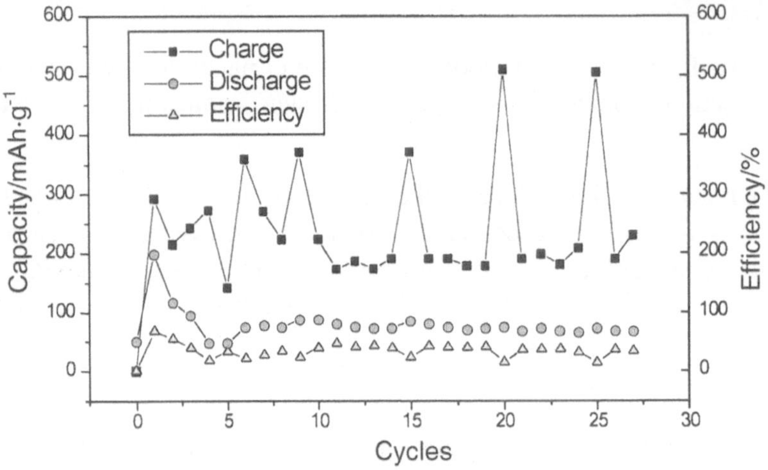

Fig. 2. Specific charge/discharge capacity and efficiency *vs.* cycle number for a small-sized hydrogen container as cathode in KOH electrolyte and NiOOH as anode

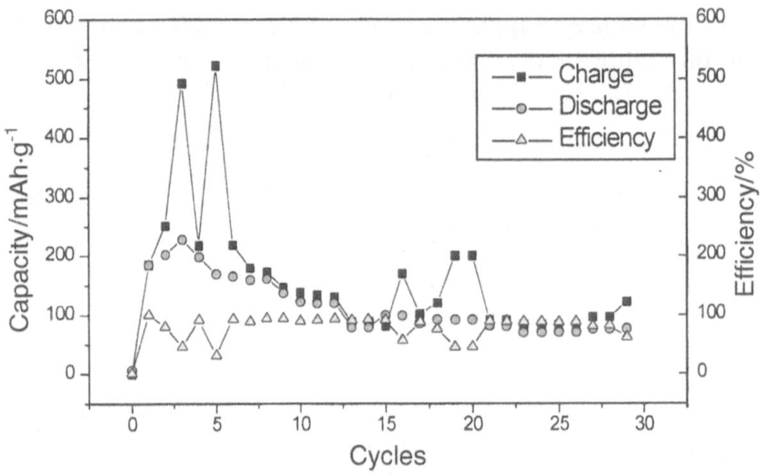

Fig. 3. Specific charge/discharge capacity and efficiency *vs.* cycle number for a large-sized hydrogen container as cathode in KOH electrolyte and NiOOH as anode

cycled upto 30 times in a laboratory 4-electrode glass cell at a current density of 74 mA/g (smaller hydrogen container) and 75 mA/g (larger hydrogen container).

Our experimental results shown in Figs. 2 and 3 indicate that the mechanism of the reactions on the *Me*H nanoparticle (and hydrogen storage nanotubes or containers) electrode surfaces in OH^- solutions (6 *M* KOH electrolyte) can be represented by Eqs. (4), (5), and (6). Between the two types of hydrogen containers, no difference can be detected in the charge/discharge cycling procedure in alkaline electrolyte.

Experiments with non-aqueous electrolyte (1 M $LiClO_4$)

We decided to find an alternative way to solve the thermodynamic and experimental problems encountered with water and aqueous electrolytes in *Me*H batteries. The estimated standard free energy formation of a compound between lithium and hydrogen atoms makes the following reaction feasible:

$$\mathrm{Li} + \mathrm{H} \rightarrow \mathrm{LiH} \tag{9}$$

The reaction according to Eq. (9) shown in Fig. 1 is more preferable throughout the whole temperature range with respect to reactions according to Eqs. (5) and (6). Two more reactions (Eqs. (10) and (11)) of charge/discharge in the Li^+ electrolyte with regard to hydrogen containers may take place:

$$\mathrm{Li}^+ + \mathrm{e}^- \rightarrow \mathrm{Li} + \mathrm{H} \Rightarrow \mathrm{LiH} \tag{10}$$

$$\mathrm{LiH} - \mathrm{e}^- \rightarrow \mathrm{Li}^+ + \mathrm{H} \tag{11}$$

The samples were prepared according to Ref. [2] as experimental electrodes by mixing nanoscaled particles (and containers) with teflonized acetylene black (TAB-2) as a cathodic mixture in a weight ratio of 50:50 and Li metal as anodic material. A Li/Li^+ reference electrode and a non-aqueous electrolyte (*1 M* $LiClO_4$ in *PC* + *EC* (propylene carbonate : ethylene carbonate = 1:1, Fluka®)) were employed. The electrodes were separated by a glass fiber paper (Amer-Sill). The cathodes were prepared by pressing the cathode mixture onto Ni or Al foil disc collectors (15 mm). The probes were cycled up to 60 times as cathodes in special stainless lab cells fitted with a steel spring in order to maintain a definite pressure ($2.5\,\mathrm{kg} \cdot \mathrm{cm}^{-2}$) on the electrode face and using $LiClO_4$ as electrolyte with a moisture content of up to 20 ppm. The cells were charged and discharged at a constant current of $74\,\mathrm{mA} \cdot \mathrm{g}^{-1}$ of the active material (*cd*: $0.5\,\mathrm{mA} \cdot \mathrm{cm}^{-2}$) to an upper limiting voltage of 4.20 V. When the limiting discharge voltage of 1.00 V was reached, the next cycle was started. The experimental results for both types of containers are shown in Table 2.

In the case of Ni foil, the smaller container working as a cathode had an initial discharge capacity. For the larger container, no initial discharge capacity could be detected experimentally. For the case of

Table 2. Specific charge (Ch.)/discharge (Disch.) capacity and efficiency (Eff.) *vs.* cycle number for small- and large-sized hydrogen containers as cathodes in $LiClO_4$ electrolyte

Number	Ni foil						Al foil					
of cycling	Small container			Large container			Small container			Large container		
	Capacity/mAh · g^{-1}						Capacity/mAh · g^{-1}					
	Ch.	Disch.	Eff./%	Ch.	Disch.	Eff./%	Ch.	Disch.	Eff./%	Ch.	Disch.	Eff./%
0	0	191	0	0	0	0	0	0	0	0	377	0
5	208	207	99	226	224	99	205	201	98	222	205	93
10	195	187	96	199	198	99	179	179	100	183	168	92
15	165	164	99	179	178	99	160	169	95	170	153	90
20	150	150	100	166	163	98	147	148	99	155	148	95
25	119	119	100	156	153	98	128	128	100	303	185	76
30	115	117	98	140	137	98	121	120	99	161	155	97
35	116	116	100	–	–	–	113	112	99	148	143	97
40	99	99	100	–	–	–	100	100	100	155	155	100
45	95	95	95	–	–	–	–	–	–	155	158	98
50	85	85	100	–	–	–	–	–	–	135	143	94
55	–	–	–	–	–	–	–	–	–	131	104	80
60	–	–	–	–	–	–	–	–	–	84	124	68
65	–	–	–	–	–	–	–	–	–	128	124	97

Al foils, contrary to Ni collector, the smaller container working as a cathode had no initial discharge capacity, whereas an initial discharge capacity was detected for the larger container. The efficiency of the charge/discharge capacity in both cases (Ni and Al foils) is more constant for the smaller container. However, a comparison of the specific capacities of the two samples (Al foils) at the 30th cycle shows very similar or equal values as those obtained using Ni foils at the 25th cycle (Table 2). At the 30th cycle, the determined capacities for larger and smaller containers are close to 155 and 120 mAh $\cdot$ g^{-1}, respectively.

Working with $LiClO_4$ electrolyte, a well-determined difference between the specific capacity of the larger nanoscaled hydrogen container in comparison to the capacity of the smaller nanoscaled hydrogen container was observed and estimated. The specific capacities are proportional to the respective specific surface areas of the containers, not to the content of hydrogen (in weight percents) or to the hydrogen density [4, 9, 10].

Acknowledgements

The authors want to express their gratitude to the management committee of COST 523 and personally to Prof. *H. Hofmann* for accepting Bulgaria among the member countries of this COST action and for contributions to the work of the Mid-Term Workshop held in Limerick, Ireland (October 2–6, 2001).

References

[1] Dragieva I, Mladenov M, Popov A, Stoynov Z (2001) In: Program of 22nd International Power Sources Symposium, Manchester, England, April 9–11, p 15
[2] Dragieva I, Deleva Ch, Mladenov M, Zlatilova P (2001) Lecture at NATO New Trends in Intercalation Compounds for Energy Storage, NATO Advanced Study Institute, Sept. 21–Oct. 2, Sozopol, Bulgaria (accepted for publication)
[3] Dragieva I, Klabunde K, Stoynov Z (2001) J Scripta Mater **44**: 2187
[4] Mehandjiev D, Dragieva I, Slavcheva M (1985) J Magn Magn Mat **50**: 205
[5] Dragieva I, Buchkov D, Mehandjiev D, Slavcheva M (1988) J Magn Magn Mat **72**: 109
[6] Mehandjiev D, Dragieva I, Slavcheva M, Buchkov D (1989) Comm Dep Chem **22**: 338
[7] Dragieva I, Mehandjiev D, Slavcheva M (1990) J Magn Magn Mat **83**: 460
[8] Dragieva I, Slavcheva M, Buchkov D, Mehandjiev D (1990) J Magn Magn Mat **89**: 75
[9] Dragieva I (1990) In: Emin D et al (eds) Boron-Rich Solids, AIP Conference Proceedings 231. Amer Inst Phys, Albuquerque, NM, p 516
[10] Nikolov S, Dragieva I, Buchkov D (1990) In: Emin D et al (eds) Boron-Rich Solids, AIP Conference Proceedings 231. Amer Inst Phys, Albuquerque, NM, p 294
[11] Mehandjiev D, Dragieva I (1991) J Magn Magn Mat **101**: 167
[12] Dragieva I (1999) In: Nedkov I, Ausloos M (eds) Nano-Crystalline and Thin Film Magnetic Oxides, NATO Science Series, 3. High Technology, 72, Kluwer, p 165
[13] Mitov M, Popov A, Dragieva I (1999) J Appl Electrochem **29**: 59
[14] Bliznakov S, Ivanova G, Dragieva I, Popov A, Stoynov Z (2000) In: Julien C, Stoynov Z (eds) NATO Science Series 3. High Technology, 85, Kluwer p 619
[15] Dragieva I, Slavcheva M, Buchkov D (1986) J Less Common Met **117**: 311
[16] Lide DR et al (eds) (1997–1998) Handbook of Chemistry and Physics, 78th edn. CRC Press, Boca Raton, New York, pp 9–55

Received October 25, 2001. Accepted (revised) January 3, 2002

Magnetic Properties of Ball-Milled Nanocrystalline Alloys $Fe_{78}B_{13}Si_9$

Marek Pekala[1,*], **Jan Grabski**[2], and **Martyna Jachimowicz**[3]

[1] Chemistry Department, Warsaw University, PL-02089 Warsaw, Poland
[2] Physics Department, Warsaw University of Technology, PL-00662 Warsaw, Poland
[3] Department of Materials Science and Engineering, Warsaw University of Technology, PL-02524 Warsaw, Poland

Summary. Magnetic properties of nanocrystalline $Fe_{78}B_{13}Si_9$ alloys are studied for three series prepared by ball milling starting from amorphous ribbons, crystallized ribbons, and elemental powders. Temperature variation of static magnetization results in strong ferromagnetic interaction which is weakly dependent on the initial material. Magnetic hysteresis loops show that saturation magnetization, magnetic remanence, and coercive field increase with frequency for both series of ribbon samples, whereas they decrease for alloys prepared from elemental powders. Power losses raise faster for the alloys prepared from elemental powders than for the two other alloys.

Keywords. Nanostructures; Alloys; Magnetic properties; Solid state synthesis.

Introduction

Nanotechnology offers new routes to synthesize new materials as well as to modify materials of classical chemical composition. The unique properties of such materials are due to a specific internal structure composed of grains of nanometer size. A remarkable amount of atoms is located at grain surfaces; their local structural and chemical coordination may be different from those it the grain cores. Depending on the way of synthesis, up to 50% of the atoms may occupy such surface or grain boundary sites. This in turn determines physical properties of the nanomaterials [1].

The above considerations raise the important question how properties are affected by the initial state of the materials from which the nanoproducts are prepared, especially when the ball milling technique is used. In order to compare the properties of final nanomaterials, $Fe_{78}B_{13}Si_9$ alloys were selected and processed in various ways. Amorphous alloys of the same composition have been studied several years ago as potential transformer cores [2–4]. As the applicability range is determined by magnetization, coercivity, and power loss properties, a response of these alloys to magnetic fields is studied at various frequencies. The observed hysteresis arises from the fact that the non-equilibrium magnetic system in alloys is moved around in phase space by an external magnetic field [5].

* Corresponding author. E-mail: pekala@chem.uw.edu.pl

Results and Discussion

The experimental results show that high-energy ball milling produces a nanocrystalline structure for all starting materials studied. The average final crystallite sizes are about 8 to 16 nm. Structural evolution of alloys occuring during the milling process affects the magnetic properties of the alloys.

All alloys studied exhibit a strong ferromagnetic ordering, independent of the starting material. The room temperature values of static magnetization of the *K* and *P* alloys are in the range of 180 to 189 $A \cdot m^2/kg$ depending on milling time. Typical magnetization curves of *P* alloys are shown in Fig. 1. The *A* alloys exhibit a somewhat smaller room temperature magnetization of 180 $A \cdot m^2/kg$. The *Curie* temperature of the nanocrystalline phases determined from temperature variation of magnetization amounts to 1050 K and confirms strong magnetic interactions. Temperature variation of magnetization allows to detect the presence of various magnetic phases as well as their respective *Curie* temperatures. The magnetic hysteresis loops registered at room temperature show a smooth evolution of magnetization, coercivity, and power losses with milling time.

Saturation magnetization of the *K* alloys is spread up to 0.18 T depending on frequency and increases with milling time (Fig. 2). For these alloys containing a mixture of α-Fe(Si) and Fe_2B phases, the parallel increase of magnetic remanence and coercive field with milling time can be seen from Figs. 3 and 4. For the longest milling time of 450 h, when the alloys attain the nanocrystalline structure, saturation magnetization and coercive field increase approximately twice upon varying the frequency from 50 to 50000 Hz (Figs. 2 and 4). Figure 5 shows that also the power losses increase abruptly with milling time. The coercive fields and power

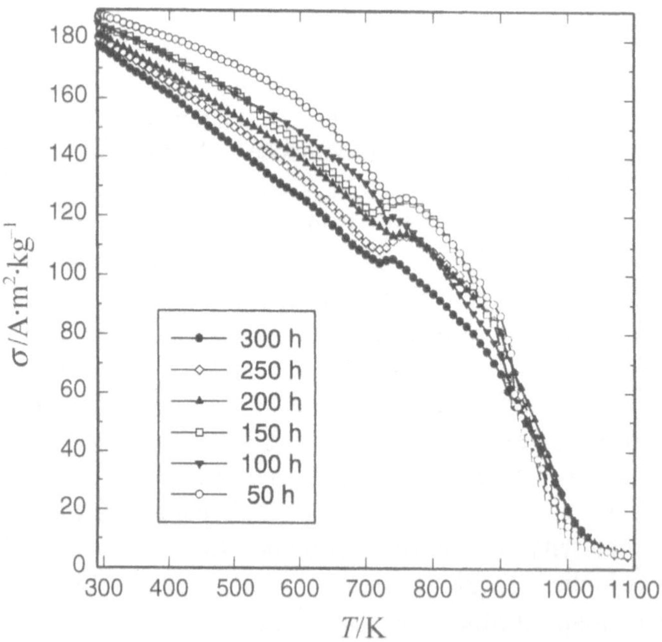

Fig. 1. Temperature variation of mass magnetization for $Fe_{78}B_{13}Si_9$ alloys *P* prepared from a mixture of elemental powders at various milling times

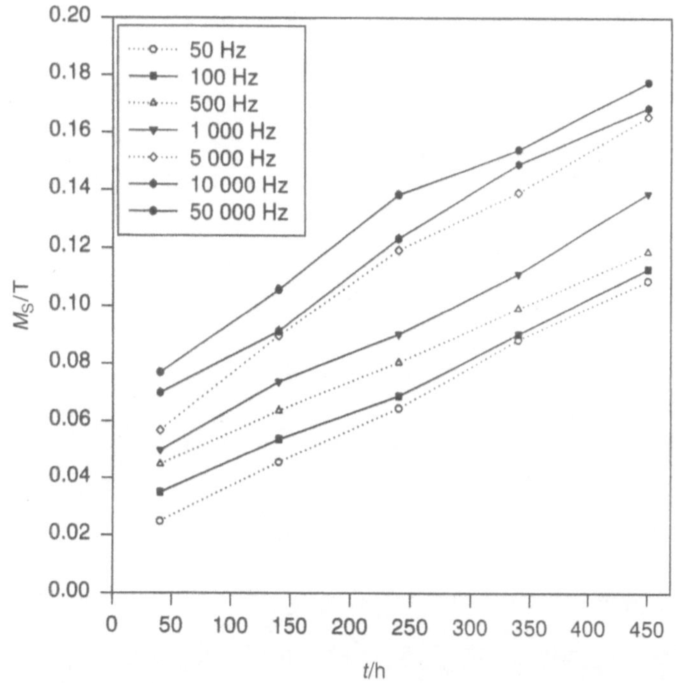

Fig. 2. Variation of saturation magnetization with milling time at various frequencies for alloys *K* prepared from crystallized ribbons

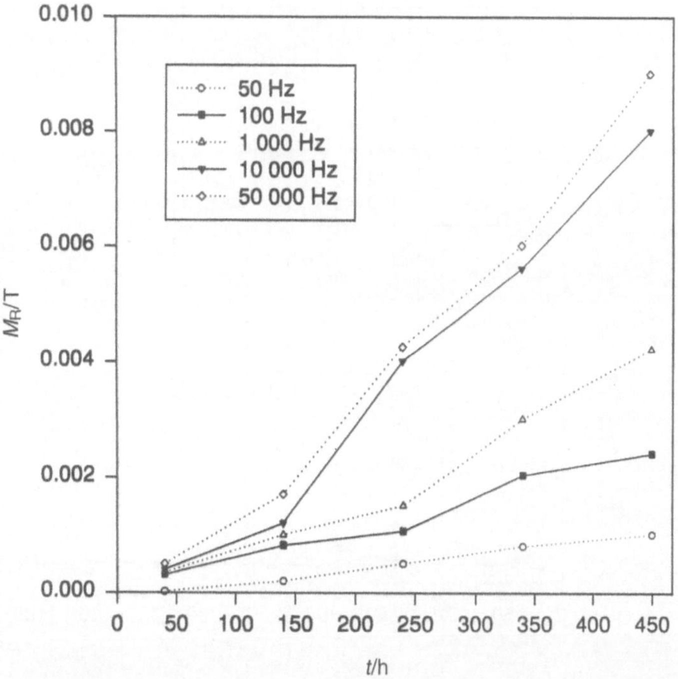

Fig. 3. Variation of magnetic remanence with milling time at various frequencies for alloys *K* prepared from crystallized ribbons

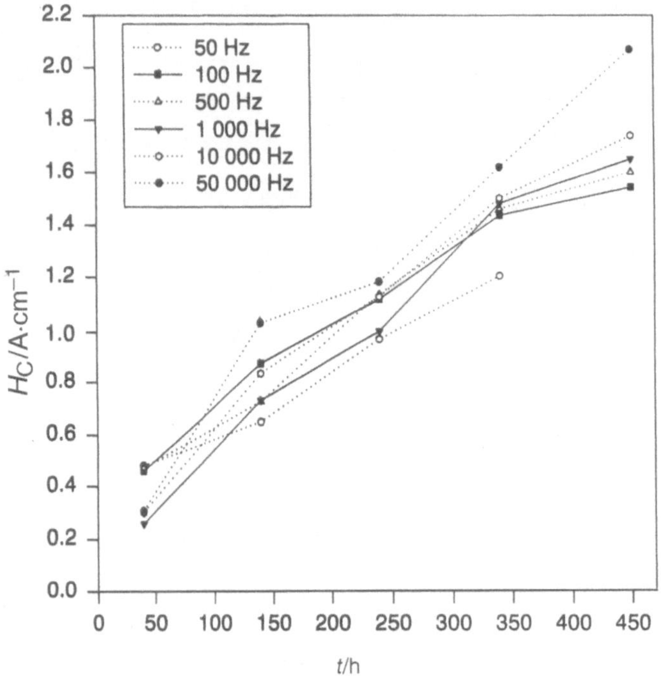

Fig. 4. Variation of coercive field with milling time at various frequencies for alloys *K* prepared from crystallized ribbons

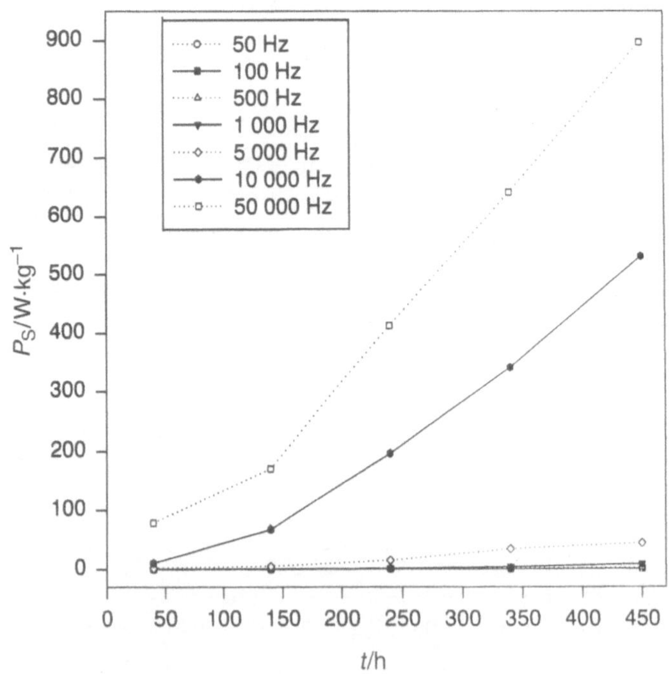

Fig. 5. Variation of power losses with milling time at various frequencies for alloys *K* prepared from crystallized ribbons

losses increase also for *P* alloys; in these alloys, the Si atoms are dissoluted, and the B atoms migrate into the iron lattice.

The above described behaviour is in contrast to that of *A* alloys, in which saturation magnetization, magnetic remanence, and coercive field diminish with milling time. On the other hand, power losses increase with milling time. Such a behaviour is related to a decay of the amorphous phase. The nanocrystalline structure made up from α-Fe(Si) and Fe_2B phases develops during the milling process. This transformation from the amorphous to the nanocrystalline phase has been confirmed by magnetization curves and *Moessbauer* spectroscopy [7].

The oscillating magnetic field strongly influences coercivity and magnetic losses in the alloys studied. Saturation magnetization and magnetic remanence increase monotonically with frequency for *A* and *K* alloys (Figs. 6 and 7), whereas these parameters diminish for the *P* alloys. Coercivity increases with frequency in *A* and *K* alloys as shown in Fig. 8. The same tendency has been reported for amorphous Fe-based alloys [8]. Figure 8 also shows that the coercive field of the *P* alloys decreases with increasing frequency.

Magnetic losses P_S raise with frequency as shown in Fig. 9 for the three alloy series. Absolute values of P_S are of the same order of magnitude in the *A* and *K* alloys when compared at the same frequency and agree with values reported for $Fe_{78}B_{13}Si_9$ alloys [9–13]. At lower frequencies, values of P_S are about one order of magnitude smaller for the *P* alloys.

In soft magnetic materials, magnetic losses are composed of two contributions. The first one is related to hysteresis losses occuring due to reordering of magnetic domain structures in an oscillating magnetic field and is approximately

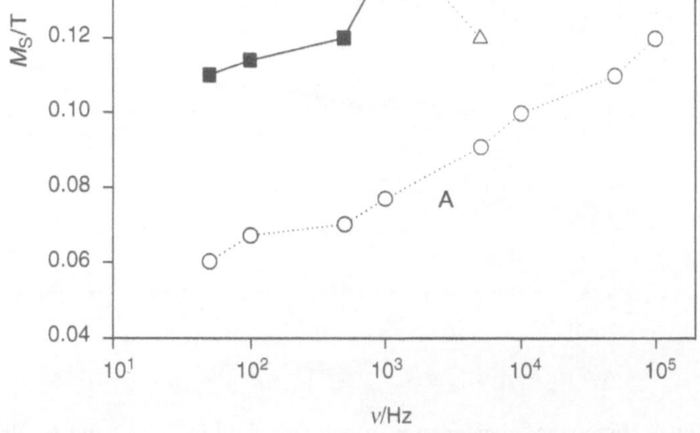

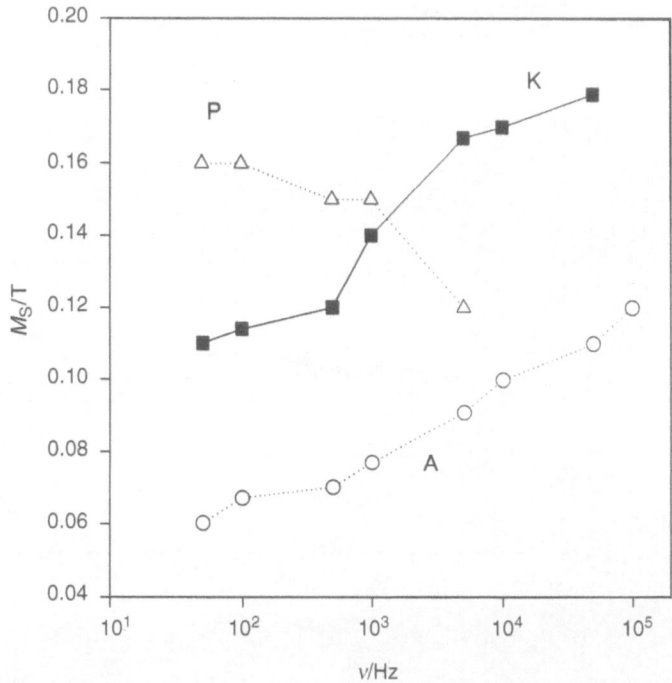

Fig. 6. Frequency dependence of magnetization for *A*, *K*, and *P* alloys

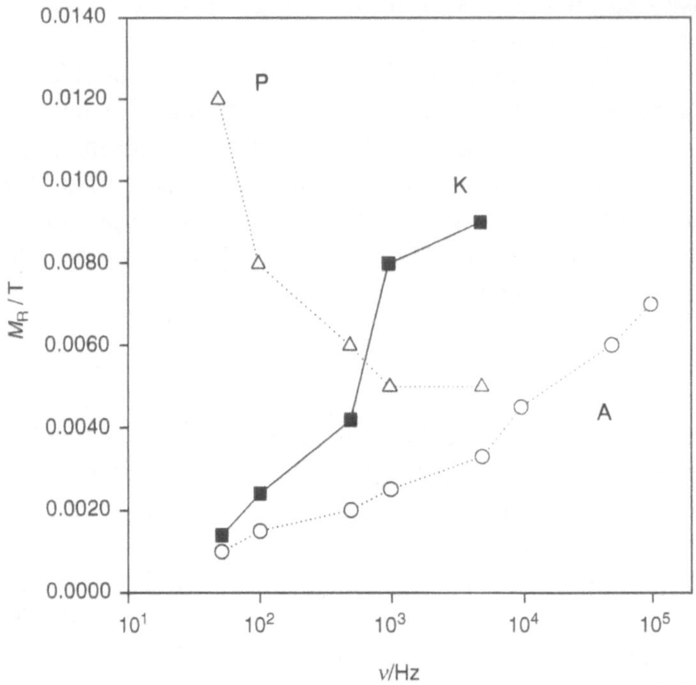

Fig. 7. Frequency dependence of magnetic remanence for *A*, *K*, and *P* alloys

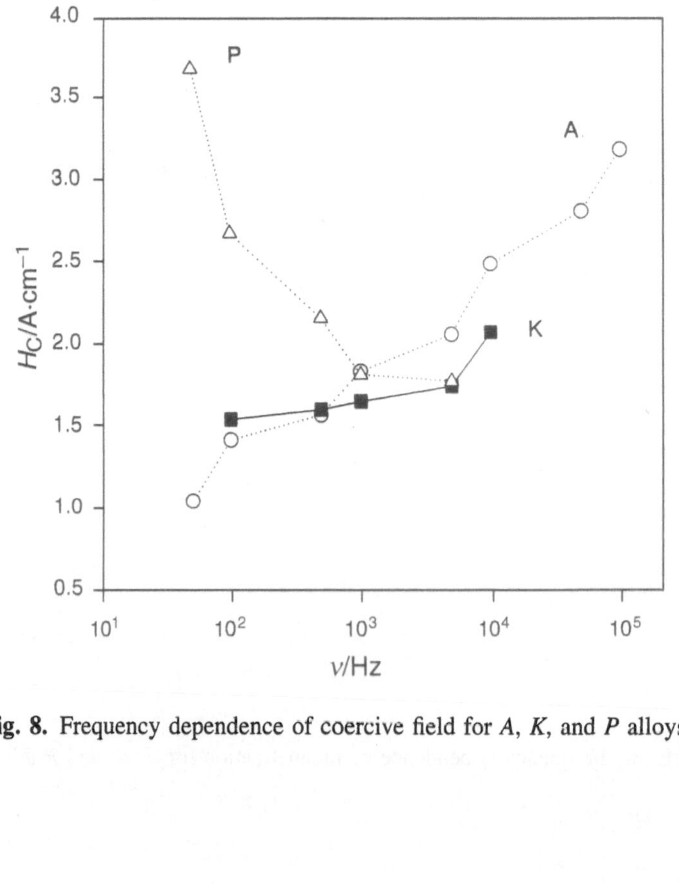

Fig. 8. Frequency dependence of coercive field for *A*, *K*, and *P* alloys

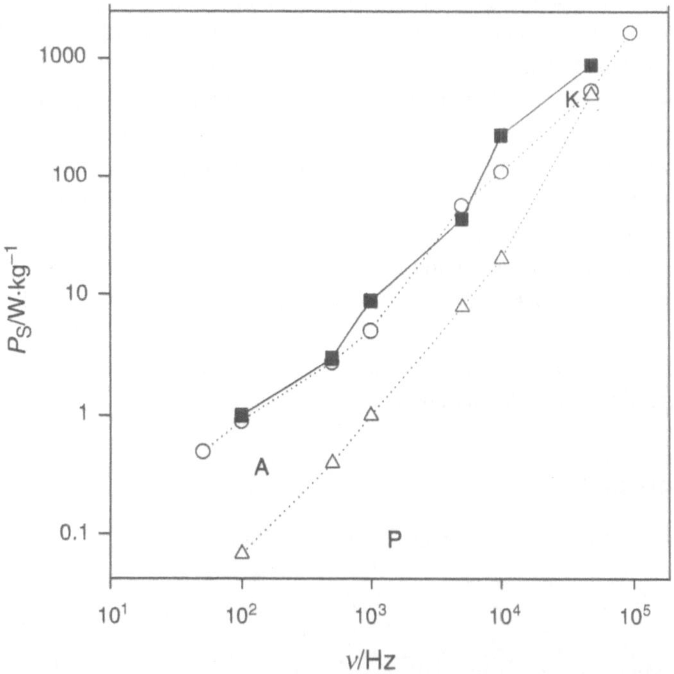

Fig. 9. Frequency dependence of power losses for *A*, *K*, and *P* alloys

proportional to the frequency ν [14, 15]. For $Fe_{74.5}Cu_{0.75}Nb_{2.25}Si_{13.5}B_9$, losses have been approximated by ν^z, z ranging between 1.3 and 1.8 [16]. The second contribution, related to eddy current losses, is proportional to ν^2 [9, 14] and dominates at higher frequencies. Plots of power losses *vs.* frequency allow to calculate the exponents governing these dependences. The values of z are very close to 1 for the *A* and *K* alloys, whereas for *P* alloys this exponent amounts to about 1.3.

Frequency variation of magnetic loop parameters is related to the rate dependent hysteresis of the magnetic system, which is driven both by the external magnetic field and thermal excitations. A response (accommodation) of the magnetic system to the actual magnetic field depends on the height of energy barriers and the rate of thermal relaxation occuring at given frequency.

The ball milling route was found to be an effective way to produce nanocrystalline alloys with properties required for practical applications; the distribution of magnetic parameters depends only weakly on the form of starting materials used for the synthesis of the alloys.

Experimental

The ball milling procedures were applied to three materials of the composition $Fe_{78}B_{13}Si_9$: amorphous ribbons, crystallized ribbons, and a mixture of elemental crystalline powders. Amorphous ribbons with a thickness about 0.02 mm produced by the melt spinning technique were cut into pieces of about 5×5 mm. Part of the ribbon was fully crystallized by annealing at 900 K for 1 h. Both the amorphous and crystallized ribbons were initially broken and used as starting materials for the ball milling procedure, which produced the '*A*' and '*K*' alloys, respectively. As a third starting material, a mixture of crystalline powders of desired atomic composition, resulted in the '*P*' alloys. The ball milling was performed in a vibratory mill under Ar. Stainless steel vials and balls were used. The weight ratio of balls to milled material was 5:1. During milling, small quantities of powder were removed from the vial in a glove-box filled with Ar. More details of the preparation are described in Refs. [6, 7].

X-Ray diffraction (XRD) and differential scanning calorimetry (DSC) were used for alloy characterization. Contamination mainly by oxygen and iron originating from the milling process did not exceed 2 at.%. The lattice parameters were determined from the diffraction patterns by least square fitting with employing an angle correction function. Mean crystallite size D and average microstrain were calculated using the *Cauchy-Gauss* method with errors not exceeding 10 and 15%, respectively. Static magnetization measurements were made with a *Faraday* balance with an accuracy of below 1%. Magnetic hysteresis loops were recorded over a broad frequency range (50 to 400 000 Hz) at magnetic field amplitudes of 500 and 5000 A/m.

Acknowledgments

This work was supported in part by grant No. 120-501/SPUB-77.

References

[1] Inoue A (2000) Acta Mater **48**: 279
[2] Kunitomo N (1994) Mater Sci Eng **A181–182**: 1296
[3] Dmowski R, Puzniak R (1984) Acta Magn (PL) Suppl 244
[4] Allia P, Baricco M, Vinai F (1990) J Magn Magn Mat **83**: 347
[5] Bertotti G, Basso C, Beatrice C, LoBue M, Magni A, Tiberto P (2001) J Magn Magn Mat **226–230**: 1206
[6] Pekala M, Jachimowicz M, Fadeeva VI, Matyja H, Grabias A (2001) J Non-Crystalline Solids **287**: 380
[7] Pekala M, Jachimowicz M, Fadeeva VI, Matyja H (2001) J Non-Crystalline Solids **287**: 360
[8] Zhukov A, Vazquez M, Velazquez J, Garcia C, Valenzuela R, Ponomarev B (1997) Mater Sci Eng **A226–228**: 753
[9] Warlimont H (2001) Mater Sci Eng **A304–306**: 61
[10] Yoshizawa Y, Oguma S, Yamauchi K (1988) J Appl Phys **64**: 6044
[11] Makino A, Suzuki K, Inoue A, Hirotsu Y, Masumoto T (1994) J Magn Magn Mat **133**: 329
[12] Makino A, Bitoh T, Kojima A, Inoue A, Masumoto T (2001) Mater Sci Eng **A304–306**: 1083
[13] del Real RP, Prados C, Pulido E, Hernando A (1993) J Appl Phys **73**: 6618
[14] Luborsky FE (1980) In: Wohlfarth EP (ed) Ferromagnetic Materials, vol 1. North-Holland, Amsterdam, p 451
[15] McHenry ME, Willard MA, Laughlin DE (1999) Progress in Mater Sci **44**: 291
[16] Schaefer M, Dietzmann G (1994) J Magn Magn Mater **133**: 303

Received October 5, 2001. Accepted (revised) November 12, 2001

Microstructure and Magnetic Behaviour of Nanosized Fe_3O_4 Powders and Polycrystalline Films

Ivan Nedkov[1,*], **Toshka Merodiiska**[1], **Svetoslav Kolev**[1], **Kiril Krezhov**[2], **Dimitris Niarchos**[3], **Elias Moraitakis**[3], **Yoshihiro Kusano**[4], and **Jun Takada**[4]

[1] Institute of Electronics, BAS, BG-1784 Sofia, Bulgaria
[2] Institute for Nuclear Research and Nuclear Energy, BAS, BG-1784 Sofia, Bulgaria
[3] Institute of Materials Science, NCSR 'Demokritos', GR-15310 Athens, Greece
[4] Okayama University, Okayama 700-8530, Japan

Summary. The object of investigation were the magnetic interactions in nanostructured Fe_3O_4 assemblies of two kinds (powder and film) where particles of similar size present nearly uniform domains in a close to planar arrangement with spacings sufficient for magnetic interactions. We discuss the use of the soft-chemistry method, *i.e.* the modified 'ferrite plating' (MFP) technique, for the synthesis of polycrystalline films of magnetite with nanosized crystallites.

Keywords. Magnetite; Magnetic properties; Thin films; Nanostructures; Powders.

Introduction

For nanosized particles, anomalies in the physical properties may be expected; in the case of ferroxide structures they may cause changes in the superexchange magnetic interaction between Fe^{n+} ($n = 2, 3$) cations. One reason for such anomalies is the shape of the nanoparticles. It is known that in nanosized particles the formation of crystalline walls (faces), which is typical for crystals, is unlikely, and their shape is often nearly spherical. The collective magnetic behaviour of nanoscaled crystallites of sizes below the critical diameter for single-domain particles (up to 54 nm for Fe_3O_4 [1]) in the process of their solidification in a polycrystalline film is strongly related to these anomalies and is a subject of current scientific interest. Proposed investigations deal with the magnetic interactions of isolated particles (powders) of nanoscale diameter and thin films of crystallites with a similar diameter, where the intercrystallite spacing is at the limit of fundamental lengths of magnetic interactions.

In the work reported here, the object of studies was magnetite (Fe_3O_4). This compound presents a classical case of a ferrimagnetic spinel-type crystal structure

* Corresponding author. E-mail: nedkov@ie.bas.bg

and in some sense is a model system [2] for other spinel solid-state solutions. We discuss the use of the modified 'ferrite plating' (MFP) technique for the synthesis of polycrystalline films of Fe_3O_4 with nanosized crystallites. 'Ferrite plating' is a promising chemical method proposed in 1983 by *M. Abe* [3]; it is environmentally friendly and requires neither vacuum nor high temperatures. The modification of the classical technology combines ferrite plating of an aqueous suspension of nanostructured ferrite powders at temperatures of about 100°C with the ferrite coprecipitation processes developed some years ago [4]. MFP makes it possible to obtain polycrystalline thin films under very soft technological conditions, thus allowing for the preservation of the nanosize of the crystallites [5]. We also performed parallel studies on the properties of the basic nanoparticles in the initial suspension and the thin films.

Results and Discussion

The XRD, TEM, and SAED data proved that the deep black coloured powders consisted of a single Fe_3O_4 phase. The *Scherrer* formula applied to the $\langle 311 \rangle$ line and microphotographs were used to select powder samples of different average particle size as a function of aging time. Two kinds of powders with an average grain size of 38 nm ($\pm$20%; **1**) and 6.5 nm ($\pm$20%; **2**) were investigated. Grains with different shapes – spherical, ellipsoidal, and octahedral – were observed in both cases. The spherical grains content was about 80% in **2**, whereas grains with octahedral shape dominated in **1** (*ca.* 85%).

TEM and SAED (Philips CM12 operated at 120 kV) studies of films with a thickness below 1 μm confirmed unambiguously the X-ray diffraction results (Table 1) for single-phase Fe_3O_4 and a near-to-spherical shape of the crystallites with a *quasi*-random orientation and a grain size up to 10 nm (Fig. 1a). For the oxygen content in the Fe_3O_4 film, the SEM (Fig. 1a) and *Rutherford* back-scattering (Fig. 1b) data yielded a ratio Fe:O = 0.73:1 (accuracy: 4%). The slight oxygen excess in the RBS data was estimated to result from overlapping with analytical peaks originating from the SiO_2 substrate.

Figure 2a illustrates the behaviour of magnetization with respect to an applied magnetic field for powders of different particles shape. Nearly no hysteresis losses

Table 1. SAED results for some interplanar distances d_{hkl} of a 1 μm thick Fe_3O_4 film compared with X-ray JCPDS data (file 26-11136)

d_{hkl}/d_{220}	JCPDS	SEAD
111	1.633	1.638
220	1	1
311	0.8528	0.8503
222	0.8164	nor resolved
400	0.7073	0.708
422	0.5773	0.58
511	0.5444	0.547
440	0.5	0.498

a)

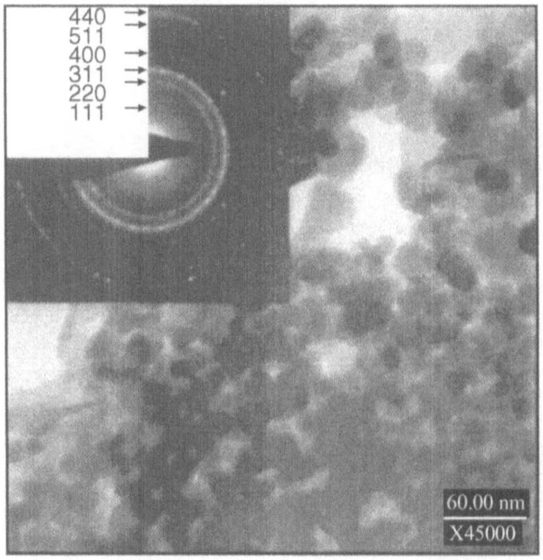

b)

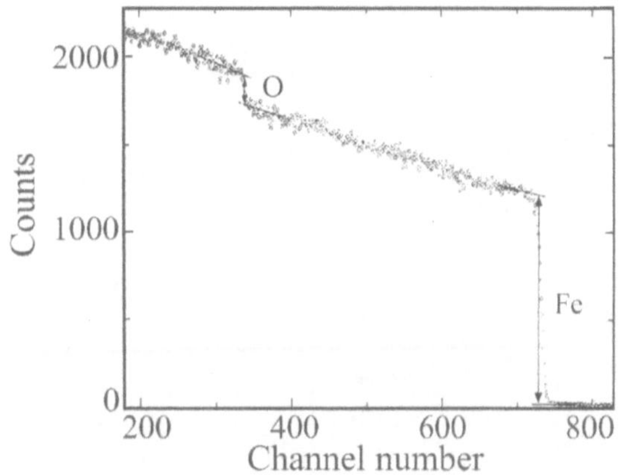

Fig. 1. a) High resolution SEM and SAED; b) *Rutherford* back scattering for the determination of the oxygen and iron content in the Fe_3O_4 thin films deposited by MFP

were observed for the two different powders. The negligible coercive field may be connected with a particle shape different from spherical (shape anisotropy) as well as with the accuracy of the VSM measurements. A considerable difference in the saturation magnetization M_S in both samples was observed. For **2**, a decrease of M_S was detected.

Figure 2b compares sample **2** and the film with 1 μm thickness. The magnetization M_{max} is rather low for the powder, whereas it is near to M_S of bulk material for the film. The hysteresis curve of the polycrystalline sample had a different shape and a slightly higher H_c (of about 150 Oe). However, in both cases of nanostructured materials (powder and film), the field dependence does not indicate clearly a *Verwey* transition.

The neutron diffraction results gave microscopic reasons for the macroscopic decrease of M_S observed for very small size particles. Using the MRIA code, a full profile refinement of the neutron diffraction pattern of a powder sample with a

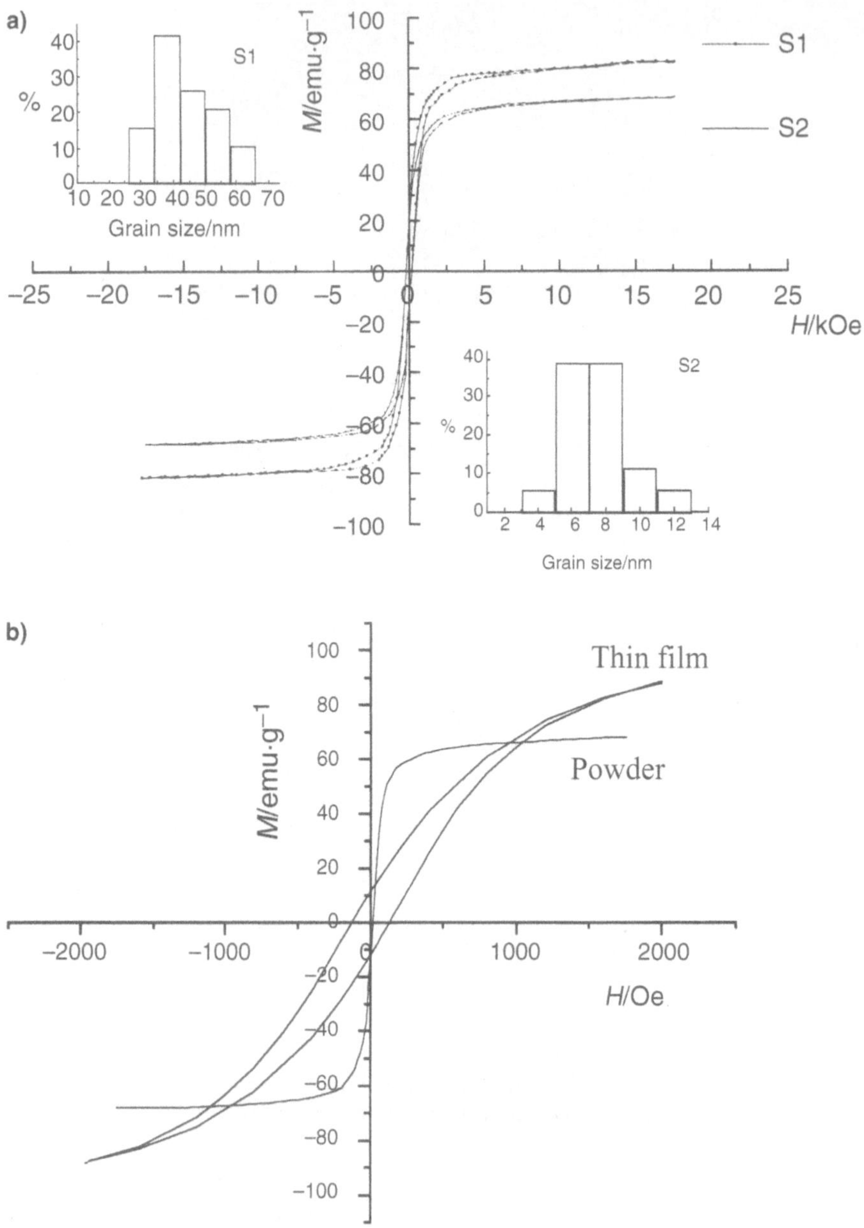

Fig. 2. a) Magnetic losses of Fe_3O_4 powders of different particle size (**1**: average particle size 38 nm; **2**: average particle size 6.5 nm) and b) *M vs. H* curves of polycrystalline films (size of crystallites: about 6.5 nm; spherical shape as obtained by MFP; room temperature)

majority of spherical particles was carried out in space group Fd-3m. Different magnetic structure models were tested. The average grain size was specified in this case as 17 ± 5 nm on the basis of a high-resolution X-ray diffraction pattern taken at room temperature and from the analysis of *Bragg* line broadening by constructing the modified *Williamson-Hall* plot [6]. For the supposed collinear ferrimagnetic structure, the rather low value of $2.6 \pm 0.3\ \mu_B$ resulted for the magnetic moment per formula unit (*M*) from the refined sublattice magnetic moments: $4.1 \pm 0.1\ \mu_B$ per tetrahedral site and $3.4\pm0.2\,\mu_B$ per octahedral site. Clearly, the estimated value of M is greatly reduced as compared to the room temperature value of $4.1\,\mu_B$ reported for bulk Fe_3O_4 (see *e.g.* Ref. [7]). Since the interparticle interaction effects are of no relevance for neutron diffraction, the reduction of magnetization reflects intrinsic properties of the particle core related with the Fe^{2+}/Fe^{3+} cationic distribution, local non-collinearity of magnetic moments, or the presence of vacancies. We have to stress at this point that modeling with assumed vacancy ordered or disordered maghemite (γ-Fe_2O_3) yielded diffracted intensity distributions which were inconsistent with the recorded neutron scattering pattern. More experimental details concerning the established dependence of the structural parameters and sublattice magnetic moments on Fe_3O_4 particle size and discussion of the possible sources for this effect have been given elsewhere [5]. It has to be noted that the lattice parameter $a_0=8.358(1)$ Å corresponding to this average particle size is significantly lower than that of bulk material. This could be related to the *Laplace* surface superpressure [8], *i.e.* the internal atoms are compressed in the spherical entities forming the nanoparticle assembly. In contrast, as mentioned above, SAED of the films yielded a lattice parameter of $a_0=8.395(1)$ Å, perfectly matching the value known for bulk material ($a_0=8.396(1)$ Å).

The analysis of the results obtained shows that MFP leads to the deposition of spherical particles only on the substrates, although the initial suspension contains particles with different shape. MFP was carried out at 97°C, and the morphology of the layers thus obtained can be explained by the difference in the dehydration of spherical and octahedral particles. Following the *Young-Laplace* equation for the pressure differential across the interface, the surface of spherical particles creates superpressure, and the internal pressure in the particle is higher than the external pressure. This could give rise to different interactions between the surface of the particle and the surrounding water as compared with the well-shaped orthorhombic particles. The particles present in the suspension with shapes different from spherical are probably 'washed away' from the film surface. The adhesion of the spherical Fe_3O_4 particles to the glass substrate is good for layers with thicknesses up to 1 μm.

Conclusions

The modified ferrite plating (MFP) technique allows the preparation of thin films containing spherical nanosized crystallites with a good adhesion to the substrate due to the dehydration process. In addition, nanostructured thin films and powders of magnetite, although consisting of nanoparticles of similar size prepared by the same technique, exhibit substantially different magnetic and crystallographic properties. The films behave like bulk material, whereas the powder assembly of nanoparticles exhibits a reduced magnetization nearly without hysteresis and a tendency for smaller lattice parameters. The difference in the interparticle interactions reflecting the different size and shape of the particles in the powder assembly as compared with the nearly uniform domains for the close-to-planar arrangement in the films might be the main reason. However, superpressure effects at an interatomic scale resulting in modified superexchange interactions in the single domain powder particles must also be taken into consideration.

Experimental

Wet chemistry was used for powder preparation. The precipitate was formed by addition of NaOH to aqueous solutions of $FeCl_2 \cdot 4H_2O$ and $NaNO_2$ mixed in a strictly fixed ratio. The precipitation process was carried out at $pH = 13$. The resulting deep black coloured substance was filtrated immediately after coprecipitation (**2**) or after aging for 24 h (**1**), washed with distilled H_2O, and dried at 50°C.

The precipitates were used to deposit Fe_3O_4 films on SiO_2 substrates by MFP [4]. The suspensions required for this purpose were prepared from the two kinds of precipitates and distilled H_2O without previous drying. The temperature of deposition was maintained at 97°C. The strongly dispersed precipitate was sprayed on the substrate with the aid of high-purity N_2. The spraying chamber was purged beforehand with N_2 for a long time to avoid oxidation by air. Substrate temperature as well as volume and velocity of the sprayed suspension were chosen to ensure the formation of a layer as uniform as possible. Variation of spraying time and suspension volume afforded layers of different thickness as measured by a *Tallystep* gauge.

X-Ray diffraction (XRD), transmission electron microscopy (TEM), and selected-area electron diffraction (SAED) were applied for phase purity analysis and characterization of the samples. The oxygen content in the Fe_3O_4 films was measured by *Rutherford* back-scattering (RBS) and scanning electron microscopy (SEM). Magnetic measurements on the initial powder and the polycrystalline films were carried out at room temperature using a vibration sample magnetometer (VSM) and a SQUID magnetometer, respectively. Neutron time-of-flight patterns of some of the powders were recorded at room temperature using the DN2 diffractometer of JINR-Dubna.

Acknowledgments

This work was supported by NATO Grant HTECH.L6.973786 and the Bulgarian National Fund for Science (contract F-816). The authors are also indebted to Dr. *L. Milenova* from BAS for her invaluable assistance in organizing some measurements.

References

[1] Aharoni A, Jakubovics JP (1988) IEEE Trans Magn **24**: 1892
[2] Kodama RH, Berkowicz AE (1999) Phys Rev **B59**: 6321
[3] Abe M, Tamaura Y (1983) Jpn J Appl Phys **2/22**: L511
[4] Jolivet P, Chaeac C, Prene P, Vayssiers L, Tronc E (1997) J Phys IV France **4/7**: C1-537
[5] Nedkov I, Merodiiska T, Koutzarova T (1998) Jpn J Magnetic Society **22**/S1: 378
[6] Somogyvári Z, Sváb E, Mészáros G, Krezhov K, Konstantinov P, Ungár T, Gubicza J (2001) Mat Sci Forum Switzerland **378–381**: 771
[7] Valenzuela R (1994) Magnetic Ceramics. Cambridge University Press, p 321
[8] Klabunde KJ (1994) Free Atom, Clusters, and Nanoscale Particles. Academic Press, San Diego, CA

Received October 22, 2001. Accepted January 21, 2002

Microstructure and Defect Characterization of Nanostructured Ni_3Al

Steven Van Petegem[1,*], **Danny Segers**[1], **Charles Dauwe**[1], **Florian dalla Torre**[2], and **Helena Van Swygenhoven**[2]

[1] Department of Subatomic and Radiation Physics, Ghent University, B-9000 Ghent, Belgium
[2] Nanocrystalline Materials Group, Paul Scherrer Institute, CH-5232 Villigen, Switzerland

Summary. Nanostructured Ni_3Al was produced by the inert gas condensation and *in situ* compaction technique and characterized by means of high-resolution transmission electron microscopy (HRTEM), X-ray diffraction, and density measurements. The defect structure was investigated using positron annihilation lifetime spectroscopy (PALS). It is shown that in some samples besides the cubic also the martensitic phase can be present. The defect structure can be divided into three major components: vacancy-like defects in the grain boundaries and nano-voids with a size of 1 nm as seen with PALS, and large pores with sizes up to 8 nm as seen with HRTEM. Furthermore, it is shown that an increasing compaction temperature leads to significantly smaller nano-voids.

Keywords. Nanostructures; Positron lifetime spectroscopy; Ni_3Al; Transmission electron microscopy.

Introduction

Ni_3Al is well known for its attractive high-temperature properties and consequent industrial applications (see *e.g.* Ref. [1]). It is also known that with a reduction of its grain size down to the nanometer scale, Ni_3Al exhibits enhanced mechanical properties like *e.g.* superplastic behaviour at low temperatures [2]. In the nano-structured regime defect structures constitute a non-negligible fraction of the solid and have therefore a big influence on its properties. In this work we present the characterization of the defect structures in nanostructured Ni_3Al produced by the inert gas condensation technique (IGC) using positron annihilation lifetime spectroscopy. Further structural characterization is done by high-resolution transmission electron microscopy (HRTEM), X-ray diffraction (XRD), and density measurements.

Positron annihilation lifetime spectroscopy (PALS) is a sensitive non-destructive tool to study defects in solids (see *e.g.* Ref. [3]). Positrons have the tendency to get trapped in open volume defects such as dislocations, vacancies, vacancy clusters, *etc.* This is due to the presence of an attractive potential at those

* Corresponding author. E-mail: steven.vanpetegem@rug.ac.be

defects. When trapped in such a defect, the lifetime of a positron increases as function of the defect size due to the lower electron density seen by the positron. Thus, probing the lifetime of a positron is equivalent to probing the size of the defects present in the solid. At large free volumes (>1 nm), positronium can be formed, which is a bound state of a positron and an electron, analogous to a hydrogen atom.

Results and Discussion

Sample characterization

Figures 1 and 2 show two typical X-ray diffraction spectra for Ni_3Al IGC samples. Sample A was compacted at room temperature, whereas sample B was compacted at 260°C. Figure 1 shows a diffraction pattern of a Ni_3Al sample with a rather 'pure' cubic structure. All three superlattice reflections {110}, {210}, and {211} show some increased intensity. The degree of ordering shows a value of 0.35, indicating 53% ordering compared to a fully ordered structure. No clear evidence can be found for the coexistence of another phase. Only one peak with a very small intensity not corresponding to the cubic phase fits to the {200} peak of the martensitic tetragonal phase of Ni_3Al. The intensity ratio between the {111} and the {200} peaks for both diffraction patterns of Figs. 1 and 2 shows a preferential orientation along the {111} planes.

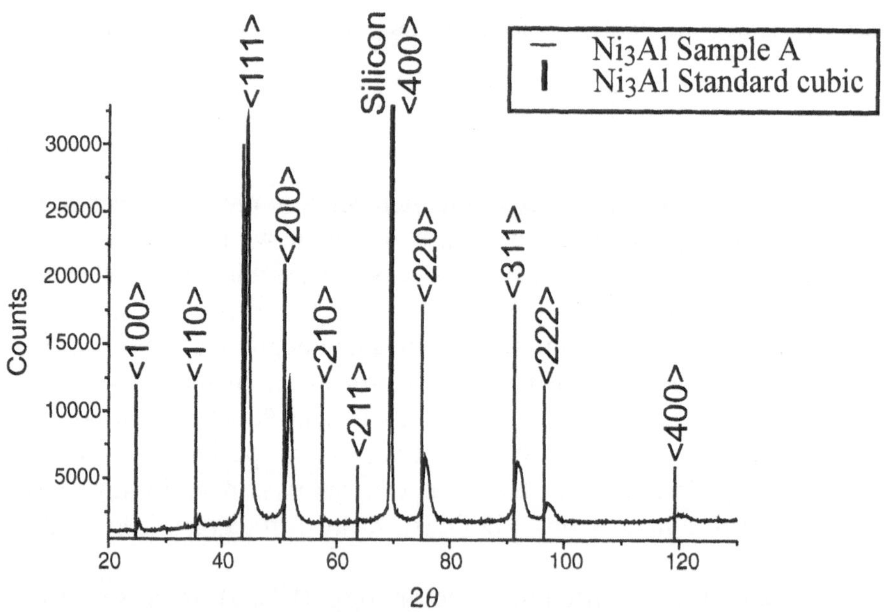

Fig. 1. X-Ray pattern of a Ni_3Al sample showing a rather 'pure' cubic microstructure; the peaks of the standard cubic Ni_3Al from the Joint Committee of Powder Diffraction Standards (JCPDS) database are added; a silicon wafer with an intense ⟨400⟩ peak at 69.4 2θ angle was added as a reference to correct for peak shifts

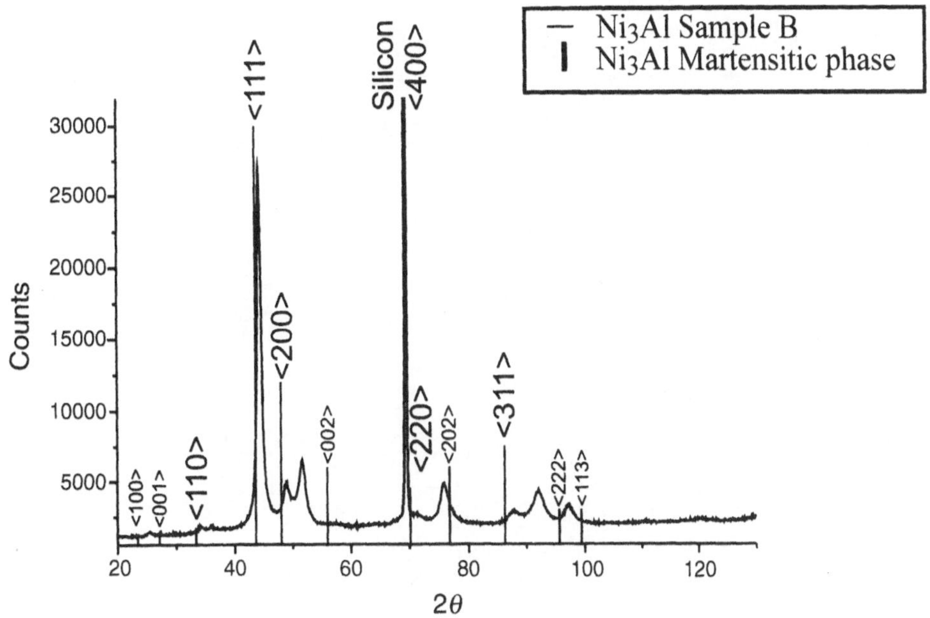

Fig. 2. X-Ray pattern of a Ni_3Al sample showing the cubic and the martensitic phase; the peaks of the standard martensitic Ni_3Al from the JCPDS database are added; the indices with small letters indicate that there is no detectable peak at this angle; a silicon wafer with an intense $\langle 400 \rangle$ peak at 69.4 2θ angle was added as a reference to correct for peak shifts

The diffraction pattern of Fig. 2 also shows ordering of the cubic structure (60% of full ordering). Additional to the peaks of the cubic phase, peaks of the martensitic phase, such as {110}, {200}, and {311} are clearly visible. The {111} peak of the martensic phase is not visible since it is superimposed on the {111} peak of the cubic phase. Further investigations showed that there is no correlation between the presence of the martensitic phase and the compaction temperature. Grain size measurements show roughly a mean grain size for sample A and B of respectively 18 and 11 nm and strains of respectively 0.558 and 0.509%. The lower grain size of sample B is probably due to the influence of the martensitic phase on the tail of the {111} peak.

In Figs. 3 and 4, HRTEM images of typical features that can be found in all samples are shown. They show a rather inhomogeneous microstructure with pores (Fig. 3) and big twins up to a size of 50 nm (Fig. 4). The untwinned grains have a mean size of about 15 nm, and grains as small as 4 nm could be observed. In Fig. 3 one can see the pores by the *Fresnel* contrast surrounding it. Pore sizes of 2 to 8 nm of elliptical shape could be found in this sample. The grains show a random orientation with mainly high-angle grain boundaries.

Corresponding to the HRTEM image of Fig. 3, a selected area electron diffraction pattern (SAED) with a selected area of approximately 500 nm can be seen in Fig. 5. The ring pattern shows the cubic phase of Ni_3Al with some ordering as seen by the reflections at {110}, {210}, and {210}. The {210} reflection is very week, indicating an incomplete ordering. No extra reflections of the martensitic phase could be found in this sample.

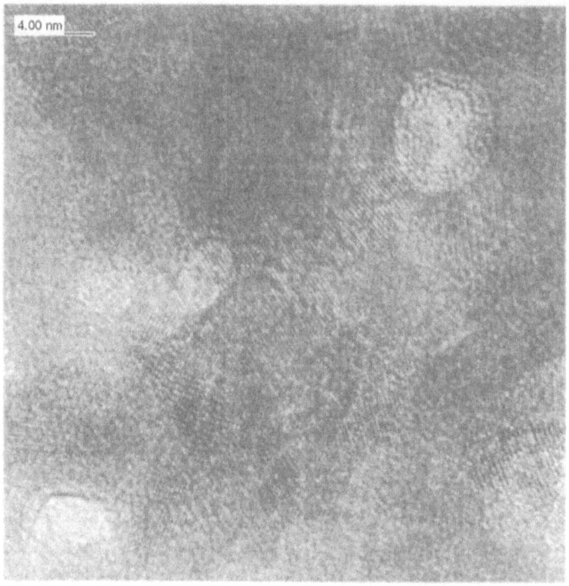

Fig. 3. HRTEM picture showing pores of about 4 to 7 nm surrounded by nanocrystalline grains

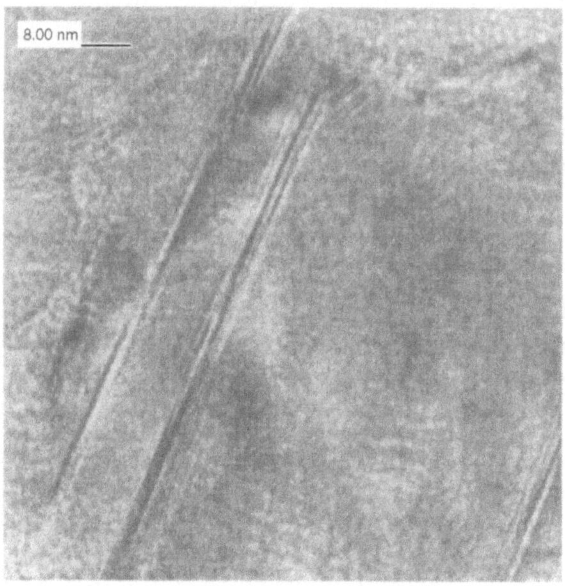

Fig. 4. HRTEM picture of twinned grains of about 50 nm

Positron lifetime spectroscopy

The positron lifetime in a well-annealed coarse-grained Ni_3Al sample was found to be 115±1 ps. This value agrees well with data found in the literature (see *e.g.* Refs. [4, 5]). This is the lifetime of free delocalized positrons in perfect crystalline Ni_3Al. In an irradiated Ni_3Al sample a component of 181±1 ps was

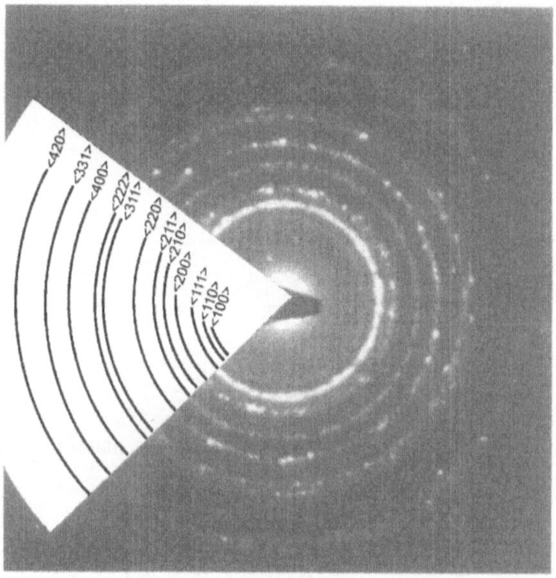

Fig. 5. SAED pattern of Ni_3Al showing some degree of ordering and no additional crystallographic phase

Table 1. Overview of positron lifetimes in nanocrystalline Ni_3Al compacted at room temperature at 2 GPa

	τ_1/ps	τ_2/ps	τ_3/ns	I_1/%	I_2/%	I_3/%
Sample 1	181±1	427±1	2.8±0.05	43±1	55±1	2.0±0.5
Sample 2	185±1	415±2	2.5±0.05	40±1	58±1	2.0±0.5

found. This is the lifetime of a vacancy in Ni_3Al. As the lifetime corresponding to a Ni and an Al vacancy only differs by a few ps [6], no distinction can be made between them.

The positron lifetime results of two nanocrystalline Ni_3Al samples are shown in Table 1. Both samples were prepared under the same conditions, *i.e.* pressed at room temperature at 2 GPa for 4 hours; their densities are 89% and 86% of that of bulk Ni_3Al. Three lifetimes could be distinguished, all of them being larger than the lifetime of free positrons in Ni_3Al. This is an indication that the defect concentration is high enough to trap all positrons into the defects. The consequence is that if more defects would be present, no change in relative intensities of the three lifetimes would be observed. This means that all information concerning the defect concentrations is lost. The shortest lifetime is comparable with the vacancy lifetime of the irradiated coarse-grained Ni_3Al sample. These defects are probably located at the grain boundaries, although one cannot exclude that some of them reside inside the grain as vacancies are stable at room temperature in Ni_3Al. In both samples a second lifetime of more than 400 ps was found. This is an indication for the existence of nano-voids. In order to estimate the size of these voids, lifetime calculations in well-known defects were performed in Ni_3Al [6]. From these

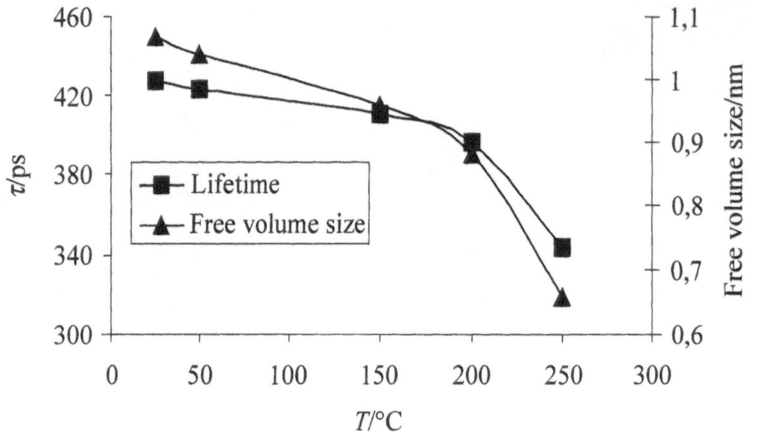

Fig. 6. Influence of the compaction temperatures on nano-void lifetime (τ_2)

calculations it was concluded that the nano-voids have a size of less than 1 nm. These small voids cannot be seen by HRTEM. The relative intensities indicate that more positrons annihilate in the nano-voids than in the vacancy-like defects. However, this is not an indication that the defect concentrations have the same ratio, as the trapping efficiency of nano-voids for positrons is much higher than of small defects. The third lifetime is due to the formation of *ortho*-positronium. No defect size estimations can be given for this component, as it is not possible to perform reliable positronium lifetime calculations in metals for the moment. This component probably originates from the pores in the sample.

In Fig. 6, the influence of the temperature during compaction on the second positron lifetime results is presented. The densities of the samples are all about 86% of that of bulk Ni_3Al. The size of the nano-voids decreases drastically to about 0.6 nm at 260°C. The lifetimes of the shortest component and of the positronium component did not change, neither did the intensities. At these temperatures, vacancies are not stable inside the grain; the shortest component therefore originates only from the grain boundary.

Conclusions

Nanocrystalline Ni_3Al produced by the inert gas condensation technique was characterised using XRD, HRTEM, and density measurements. It was shown that two phases can be present in the samples: the cubic phase, which was found in all samples, and the martensitic phase that could only be found in some samples. The HRTEM images revealed the existence of big twins (50 nm) and pores with a size up to 8 nm. Furthermore, it was shown that positron lifetime spectroscopy experiments on nanocrystalline materials can provide extra information about the defect structure that cannot be seen with other techniques. Here, three major defect structures were found: vacancy-like defects, nano-voids with a size of about 1 nm, and pores. Only the latter component could be seen using microscopy techniques. It was also shown that the compaction temperature has a major influence on the nano-voids.

Experimental

Sample preparation

A coarse grained Ni_3Al sample was produced by mixing high purity Ni and Al in a ratio of 3:1 at.% and subsequent annealing at 1600 K for 1 h. After mechanical and chemical polishing the sample was annealed at 1200 K for 100 h to obtain maximum homogeneity. The exact composition was determined as $Ni_{76}Al_{24}$. The sample was irradiated at room temperature with 2 MeV electrons at the LINAC facility [7] at the Ghent University with a dose of $1 \times 10^{18}\,e^-/cm^2$.

The nanostructured Ni_3Al samples were made by the inert gas condensation and *in situ* consolidation technique. A high-vacuum chamber (2×10^{-7} mbar) is filled with 3 mbar of high-purity (99.999%) helium. Pure cubic Ni_3Al is then evaporated from a resistively heated tungsten crucible. The evaporated atoms condense and form small clusters that are collected on a liquid nitrogen-cooled finger. The clusters are removed from the tilting cold finger with a copper scraper and then compacted under uni-axial pressure to produce disk-shaped samples of 8 mm diameter and 0.2–0.3 mm thickness. The compression unit is temperature controlled. The temperature is measured using a thermocouple close to the sample and is varied between room temperature and 260°C.

Characterization techniques

The density of the samples was measured *via* the *Archimedes* principle with a Mettler microgram balance. As reference, liquid diethyl phthalate with a density of $1.1175\,g/cm^3$ was used.

The X-ray measurements were made with a Siemens D500 X-ray detector. Applying the *Warren-Averbach* method after *Krill et al.* [8] for X-ray peak broadening, which is caused by small coherent domains, grain size and strain can be derived. Measurements were done on the $\langle 111 \rangle / \langle 222 \rangle$ family planes. A silicon wafer with an intense $\langle 400 \rangle$ peak at 69.4 2θ angle was added as a reference to correct for peak shifts. In all samples a peak shift of about 0.5° towards higher 2θ angles was observed which is due to an instrumental artefact.

HRTEM investigations were carried out at the *Centre Interdepartemental de Microscopie Electronique* (CIME) at the EPFL, Lausanne, with a Phillips CM 300 instrument operating at 300 kV with a nominal resolution of 2 Å. The specimens were first mechanically pre-thinned and then electropolished in a double-jet Tenupol 3 apparatus using a solution of 10% perchloric acid in MeOH and operating at −30°C and 13.5 V yielding an approximate current of 0.135 mA. The degree of order of the cubic phase can be determined from the ratio between the intensities of the $\{110\}$ superlattice reflection and the $\{220\}$ fundamental line.

The positron annihilation lifetimes were measured using a conventional fast–fast lifetime spectrometer (see *e.g.* Ref. [3]) with a resolution (FWHM) of about 220 ps using the sandwich arrangement. The positron source (about 0.4 MBq) was obtained by evaporating $^{22}NaCl$ onto a standard kapton foil (7 μm thickness) which was then sealed with another foil. The number of counts in each spectrum (after background subtraction) exceeded 15×10^6. The spectra were analyzed using the multi-component program LT developed by *Kansy* [9]. All spectra could be decomposed into three components with a normalized χ^2 lower than 1.1.

Acknowledgements

The authors would like to thank Prof. *Pierre Stadelman* and Dr. *Marco Cantoni* from the CIME and Dr. *Robin Schaeublin* from CRPP, all at EPFL Lausanne, for their help in operating the HRTEM. We are also grateful to Dr. *Ivonna Jirásková* for supplying the coarse grained Ni_3Al samples. This research is part of the Interuniversity Poles of Attraction Program, Belgian State, Prime Minister's Office, Federal Office for Scientific, Technical, and Cultural Affairs (IUAP 4/10).

References

[1] Liu CT, Stringer J, Mundy JN, Horton LL, Angelini P (1997) Intermetall **5**: 579
[2] McFadden SX, Mishra RS, Valiev RZ, Zhilyaev AP, Mukherjee AK (1999) Nature **398**: 684
[3] Hautojärvi P, Corbel C (1995) Positron Spectroscopy of Defects in Metals and Semiconductors. In: Dupasquier A, Mills AP Jr (eds) Positron Spectroscopy of Solids. IOS Press, Amsterdam, p 491
[4] DasGupta A, Smedskjaer LC, Legnini DG, Siegel RW (1987) Mat Sci For **15–18**: 1213
[5] Mills AP, Wilson RJ (1982) Phys Rev A **26**: 490
[6] Van Petegem S, Kuriplach J, Hou M, Zurkin EE, Segers D, Morales AL, Ettaoussi S, Dauwe C, Mondelaers W (2001) Mat Sci For **363–365**: 210
[7] Mondelaers W, Van Laere K, Goedefroot A, Van den Bossche K (1996) Nucl Instr Met A **368**: 278
[8] Krill CE, Birringer R (1998) Phil Mag A **77**: 621
[9] Kansy J (1996) Nucl Instr Met A **374**: 235

Received October 5, 2001. Accepted (revised) November 12, 2001

Characterization of Nanocomposite Coatings in the System Ti–B–N by Analytical Electron Microscopy and X-Ray Photoelectron Spectroscopy

Andreas Gupper[1], **Asunción Fernández**[2], **Christina Fernández-Ramos**[2], **Ferdinand Hofer**[1,*], **Christian Mitterer**[3], and **Peter Warbichler**[1]

[1] Forschungsinstitut für Elektronenmikroskopie, Technische Universität Graz, and Zentrum für Elektronenmikroskopie Graz, A-8010 Graz, Austria

[2] Instituto de Ciencia de Materiales de Sevilla, Centro de Investigaciones Científicas Isla de la Cartuja, E-41092 Sevilla, Spain

[3] Institut für Metallkunde und Werkstoffprüfung, Montanuniversität Leoben, A-8700 Leoben, Austria

Summary. Superhard nanocomposite coatings of different composition in the *quasi*-binary system $TiN-TiB_2$ were deposited onto stainless steel sheets by means of unbalanced DC magnetron co-sputtering using segmented TiN/TiB_2 targets. The chemistry and microstructure of a $TiB_{0.6}N_{0.7}$ coating was investigated using X-ray and electron diffraction, photoelectron spectroscopy, energy-filtering transmission electron microscopy, and electron energy-loss spectrometry. High resolution elemental mapping of the elements Ti, B, N, and O with energy-filtering TEM reveals a homogeneous distribution on the nanometer scale. X-Ray and electron diffraction exhibit only TiN crystallites of nanometer size, but no information on the boron-rich phase. The near-edge fine structures of the B*K* and N*K* ionization edges in the EELS spectra of the Ti–B–N coatings were used to derive information on the phases by comparing the edges with those of reference compounds. It was found that the TiN nanocrystals occur together with TiO_x particles; the grains are embedded in a strongly disordered or *quasi*-amorphous matrix consisting mainly of TiB_2 particles and, near the steel substrate, also boron oxide (B_2O_3).

Keywords. Nanostructures; Electron microscopy; EELS; EFTEM; XPS; XRD.

Introduction

Nanocomposite coating materials have recently attracted increasing interest due to the possibility of the synthesis of materials with unique properties, *e.g.* superhardness [1], combined high hardness and toughness [2], or hardness and low friction [3]. There is an increasing number of nanocomposite wear-resistant coating material systems, such as $TiN-Si_3N_4$ [1, 4], Ti–B–N [5], and $Ti-B-N-MoS_2$ [6].

* Corresponding author. E-mail: ferdinand.hofer@tugraz.at

In order to understand the origin of the exceptional properties of these materials and possible trends it is imperative to have a good comprehension of the multiphase composition and microstructure. The lack of knowledge in this field stems not only from the poorly understood micromechanical mechanisms, but also from the difficulty of characterizing these nanocrystalline materials, the resolution limits of analytical techniques often being not sufficient.

Previously, characterization of nanocomposite coatings has mainly been performed by X-ray photoelectron spectroscopy (XPS), *Auger* electron spectroscopy (AES), X-ray diffraction (XRD), electron-probe microanalysis (EPMA), and extended X-ray absorption fine structure analysis (EXAFS) [7]. However, since these techniques are not always sufficient for studying nanocomposite coatings, it is necessary to obtain structural information beyond that available from conventional techniques. In this paper, we apply modern techniques of transmission electron microscopy (TEM) as already demonstrated in previous investigations [8, 9]. We extend these investigations by means of a combination of an electron microscope with an imaging energy filter to perform energy-filtered imaging (EFTEM) and electron energy-loss spectroscopy (EELS). The aim of the present study is to gain chemical and structural information on the phases and to investigate the distribution of the elements with the highest possible lateral resolution. One major point of the present study is to apply the electron energy-loss near edge fine structures technique (ELNES) which has proved fruitful in studying boron and nitrogen containing compounds due to the information provided about bonding and nearest neighbours [10, 11].

Results and Discussion

In order to study nanocomposites consisting of TiB_2 and TiN, the elemental coating composition was adjusted to the *quasi*-binary tie line between the individual compounds in the system Ti–B–N [9]. The results of XRD analysis for different chemical compositions are summarized in Fig. 1. The TiB_2-based layer exhibited the well-known strong (001)-preferred orientation of the hcp TiB_2 phase [5], whereas in TiN films fcc TiN with a random orientation was formed [16]. The gracing angle traces for Ti–B–N samples did not reveal well-defined diffraction patterns as for the TiB_2- or TiN-based layers. It is already well known that the very small grain size in nanophase materials gives their diffraction pattern the appearance of an amorphous material [2, 17]. This can be also observed for the Ti–B–N samples, where the X-ray peaks are considerably broadened, thus indicating very small crystal grains in the range of some nanometers. However, it is only possible to identify the X-ray lines due to TiN crystallites. Additionally, a gradual shift of the line positions of TiN to lower angles can be observed, thus indicating a widening of the TiN lattice.

X-Ray photoelectron spectroscopy lead to some valuable information concerning the surface composition. It brought clear evidence for the existence of more than one boron and nitrogen containing compound. The B1s, N1s, and O1s spectra of one sample before and after sputtering used for interpretation are shown in Fig. 2. In the B1s spectra, two large peaks at 186.7 and 191.2 eV are visible which were assigned to TiB_2 and B_2O_3, respectively. Table 1 summarizes the binding

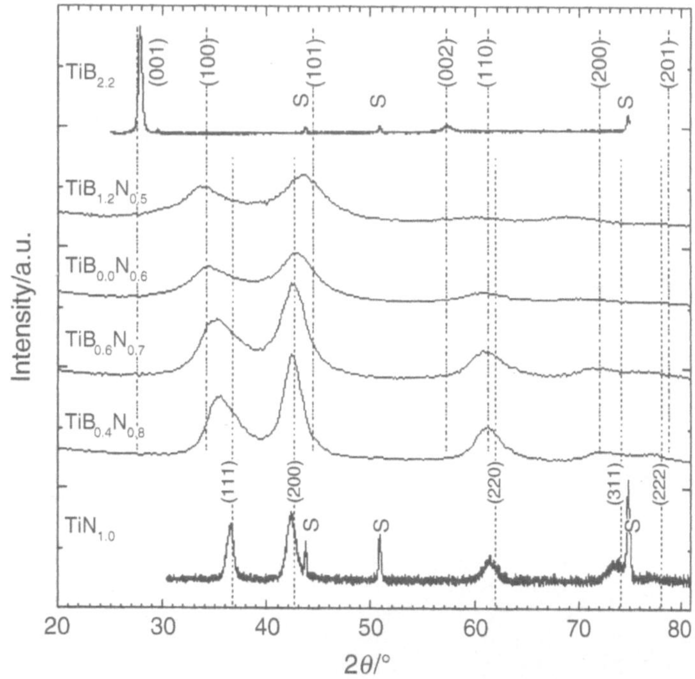

Fig. 1. XRD patterns of Ti–B–N coatings of different compositions on steel substrates

energies for possible boron containing compounds and the corresponding references. The peak for the latter component in the boron spectra appears at relatively low binding energy as compared to the literature (193.1 eV) [18]. This could be due to a deficiency of oxygen or the incorporation of titanium. Furthermore, it is obvious that there is no elemental boron present at the surface, as no signal was observed at 186.5 eV [19]. The N1s spectra show more then one signal with TiN as the main component at about 396 eV. The peak at higher binding energy disappears during the sputtering process, probably because of the removal of a nitrogen containing surface contamination. The oxygen spectrum displays two main contributions. The component at higher binding energies indicates the presence of water adsorbed onto the surface which is removed after ion sputtering. The component at lower binding energies results from the incorporation of oxygen inside the film due to the high avidity of Ti and B for this element. The existence of B_2O_3 has been already discussed, and oxidized titanium can also be seen in the Ti2p XPS spectra (not shown). The films are not only oxidized at the surface, but they present a certain level of bulk oxidation which remains after sputtering (Fig. 2) as is confirmed by quantitative EELS analysis. The quantification of the XPS spectrum for the $TiB_{0.6}N_{0.7}$ sample is shown in Table 2. The oxygen content is rather high (about 46 at%) which may be explained by surface oxidation of the nanocomposite layers, the XPS spectrum being recorded at the surface of the layer.

In case of the single-phase TiN coatings, the SEM investigation of fracture cross-sections showed the well-known columnar structures, but the layers with increasing boron content showed dense, featureless glass-like structures [8–10]. The TEM bright field image of a Ti–B–N nanocomposite cross-section foil

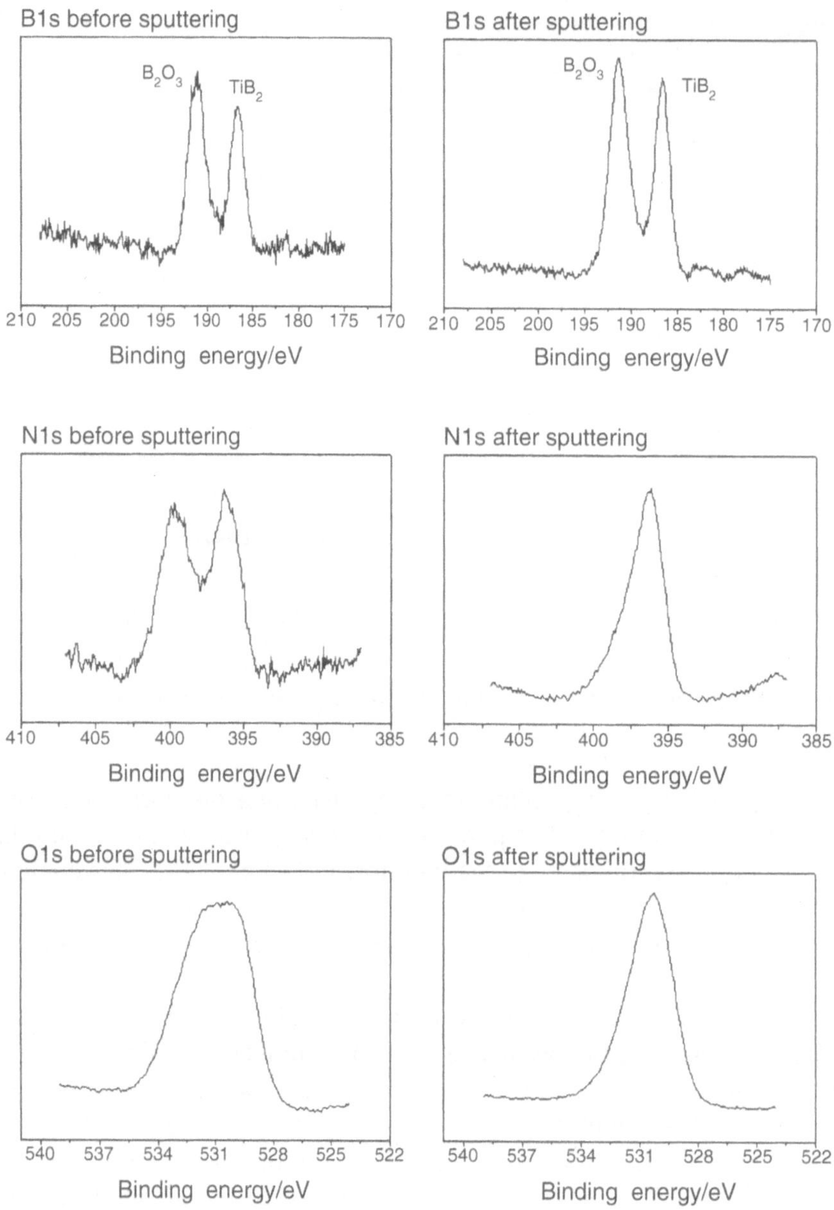

Fig. 2. XPS spectra of a coating with the composition $TiB_{0.6}N_{0.7}$; B1s, N1s, and O1s before and after sputtering

Table 1. Binding energies of some boron containing compounds

Compound	B_2O_3	BN	TiB_2	B
Binding energy/eV	193,1	190,5	187,5	186,5
Reference	[15]	[15]	[15]	[16]

Table 2. Results of the quantification of the XPS and EELS spectra of $TiB_{0.6}N_{0.7}$ (in at%) and the phases identified by different analytical techniques (XPS, XRD, ED, EELS, ELNES)

	Location	Ti	B	N	O	Compounds
XPS	Surface	25.0	21.0	7.0	46.0	
EELS	Near surface	27.8	20.6	32.2	19.4	TiB_2, TiN, TiO, C
	Middle	31.5	19.8	24.4	24.3	TiB_2, TiN, TiO, C
	Near substrate	26.0	12.0	18.4	43.6	TiB_2, TiN, B_2O_3, TiO, C

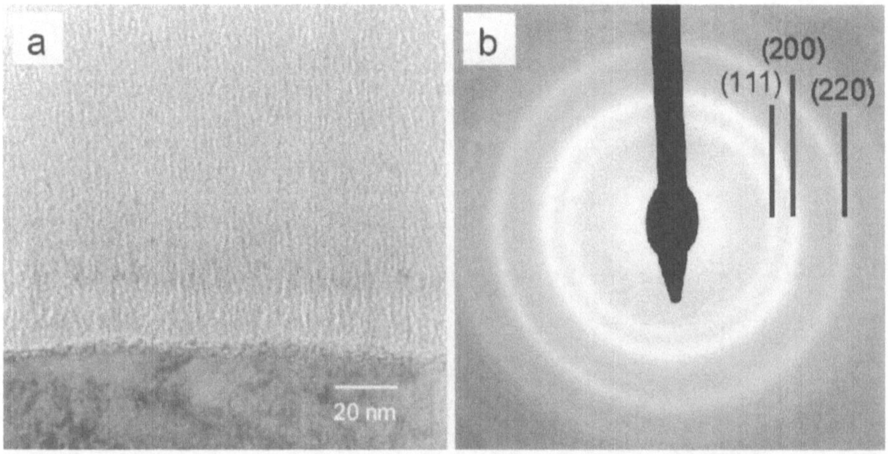

Fig. 3. TEM cross-section of a coating with the composition $TiB_{0.6}N_{0.7}$ deposited onto an austenitic stainless steel substrate; a) TEM bright field image, b) corresponding selected area diffraction pattern indicating TiN crystallites of nanometer diameter

(Fig. 3a) revealed crystallites of approximately 3 to 5 nm. With the deposition parameters used, we observed no pronounced influence on the grain size. Contrary to the well-known competitive growth and grain coarsening following nucleation and coalescence, the crystallite size does not vary significantly with increasing film thickness. The rings of the selected electron diffraction pattern shown in Fig. 3b belong to the (111), (200), and (220) reflections of the TiN phase. They are in good agreement with the XRD results obtained for the same film composition. Similar to XRD results, the measured d values are up to 6% higher than the values found for TiN. This may be explained by the incorporation of boron in the TiN phase or by a disordered structure and hence high residual stresses. However, both diffraction methods showed no sign of the TiB_2 phase, which raises the question how the boron-rich phase is distributed within the layer.

In order to investigate the phase distribution, we recorded EFTEM elemental distribution maps of the interface region near the steel substrate. Figure 4 shows the elemental maps of titanium, nitrogen, boron, and oxygen. The only features distinguishable are a thin oxygen-enriched zone at the substrate/coating interface which is probably due to the natural oxide layer on the steel surface and a thicker zone close to this interface with increased boron and nitrogen content [8].

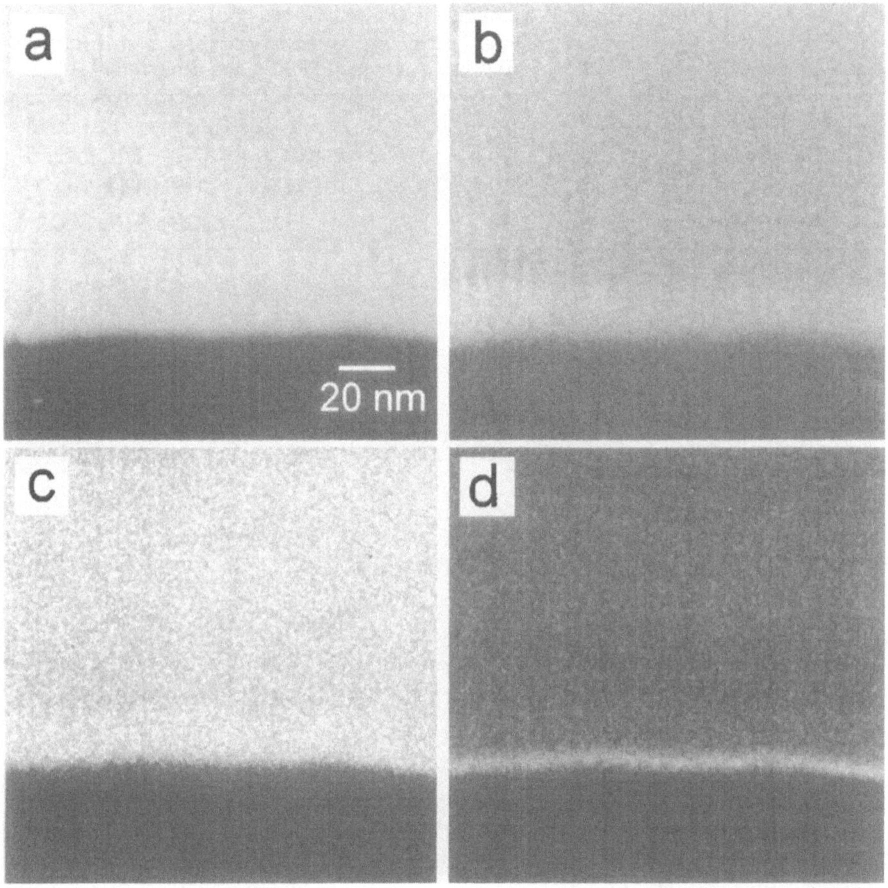

Fig. 4. EFTEM elemental maps of a coating with the average composition $TiB_{0.6}N_{0.7}$ near to the interface to the substrate; a) Ti$L_{2,3}$ jump ratio image, b) BK jump ratio image, c) NK jump ratio image, d) OK jump ratio image

Although EFTEM elemental mapping can be performed with a spatial resolution of about 1–2 nm, it is not possible to distinguish the phases present in this nanocrystalline sample. The origin of this drawback is that the TEM and EFTEM images are two-dimensional projections of three-dimensional samples. Since the specimen thickness of the ion-milled sample shown in Figs. 3 and 4 is about 30 nm and the crystals are smaller and randomly distributed, they cannot be seen in the projection. This principal limitation of EFTEM in the characterization of nanocrystalline materials has been often overlooked until now. However, in spite of the apparently homogeneous distribution of the boron and nitrogen elements in the films shown by EFTEM, the presence of two separated compounds, TiN and TiB_2, instead of a possible ternary Ti–B–N phase could be detected by the binding energy values of the corresponding XPS signals.

Nevertheless, in order to obtain more detailed information on the chemical phases of the Ti–B–N layers, we used advanced analytical techniques. In Fig. 5, EELS spectra collected for different positions on this cross-section are shown. It is evident that the coating contains not only boron, titanium, and nitrogen, but also oxygen and a small amount of carbon. Furthermore, the ionization edge intensities

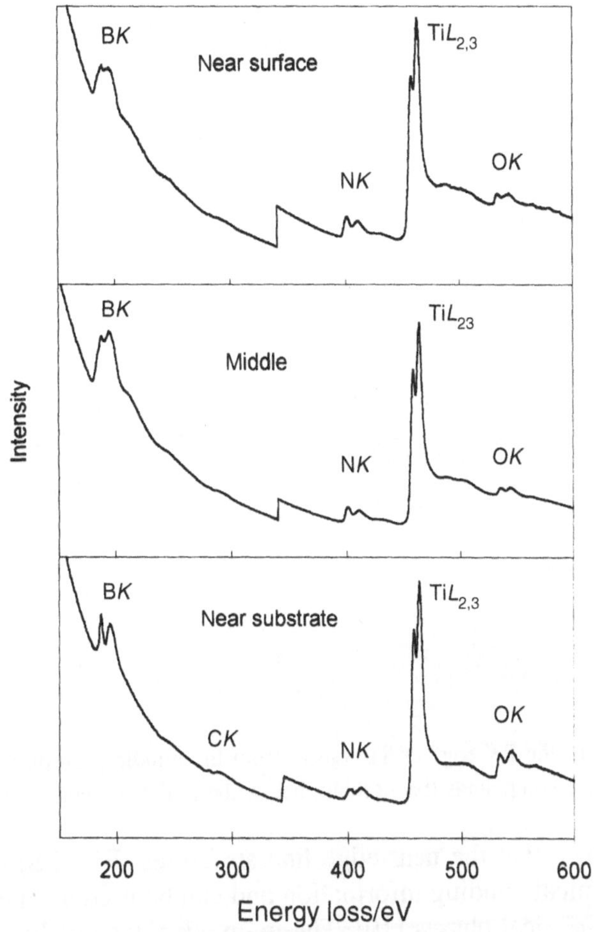

Fig. 5. EELS spectra of a coating with the average composition $TiB_{0.6}N_{0.7}$ (shown in Fig. 1) at different positions of the coating

vary due to concentration changes, *e.g.* the region near the substrate contains more oxygen but less boron, and the edge fine structures change between the different positions, thus suggesting that not only the elemental composition but also the chemical compounds vary. Particularly, the B*K* edges from different regions within the Ti–B–N layers gave a wide range of ELNES features. The spectra recorded near the steel interface exhibited similar features to those found in B_2O_3, where trigonally coordinated boron gives rise to a π^* peak at 193 eV and a σ^* peak at 204 eV [20], thus indicating that most of the boron is bound in the form of boron oxide (in the bright layer visible in the boron map, Fig. 4b). The B*K* edges recorded in the middle of the layer and near the surface showed features that are similar to the spectrum of titanium diboride (Fig. 6).

Quantification of the spectrum from the middle position gives 31.5 at% Ti, 19.8 at% B, 24.4 at% N, and 24.3 at% O; other quantification results are summarized in Table 2. The comparison of the XPS results with the EELS analysis of the surface position shows similar values for the titanium and boron concentrations, but different nitrogen and oxygen contents. This may be explained by oxidation of titanium nitride in the uppermost layers of the Ti–B–N nanocomposite, which are only probed by XPS, even after sputtering for some minutes.

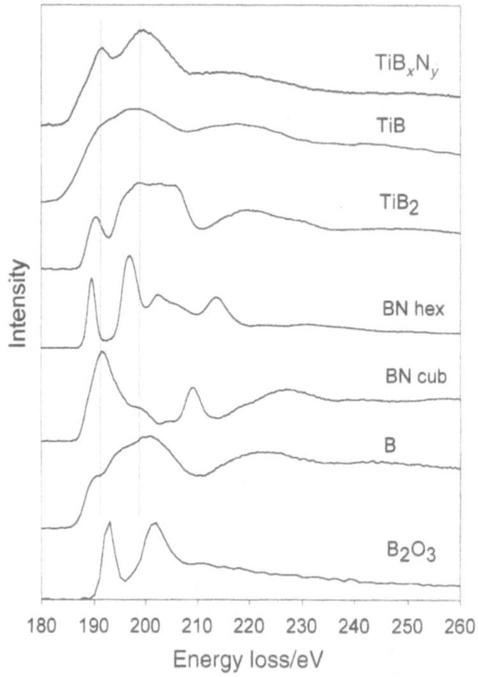

Fig. 6. Comparison of the B*K* edge of $TiB_{0.6}N_{0.7}$ from the middle position of the layer with B*K* edges from reference compounds (background below the ionization edges has been subtracted)

It is well known that the near-edge fine structures (ELNES) of the ionization edges reveal chemical bonding information and can be used as fingerprint information to identify chemical phases [10]. The main advantage of ELNES is its potential to examine changes in bonding with spatial resolution at the nanometer level and even approaching the level of interatomic changes [11].

For the elucidation of the chemical phases occurring in the Ti–B–N layers, we used the fingerprinting approach and compared the fine structures of the B*K*, N*K*, and O*K* edges recorded at the middle position of the layer ($TiB_{0.6}N_{0.7}$) with reference edges measured from TiN, TiB_2, B_2O_3, BN, and TiO_2 standards. Figure 6 shows the near-edge fine structure of the B*K* edges; the B*K* edge of the Ti–B–N coating is quite similar to the B*K* edge of TiB_2 and to some extent also to that of elemental boron and TiB. The B*K* edge of the Ti–B–N coating does not exhibit any ELNES feature of the BN compounds and of boron oxide, which means that these compounds do not occur in significant concentrations (<5 at%) in the coating. Furthermore, considering only the ELNES of the B*K* edge, the occurrence of elemental boron cannot be completely excluded, but XPS results showed that elemental boron does not occur in the Ti–B–N coating.

The main difference between the Ti–B–N coating and the TiB_2 standard is that the coating had been subjected to ion beam milling. However, it has been shown recently that ion milling does not cause a damage to nanocrystalline TiB_2 [21]. Since ELNES is strongly dependent on nearest neighbours and bonding, there is a possibility of impurity atoms causing the different spectra. A possible candidate is surely oxygen, which is measurable in all specimen regions. However, in the middle of the layer this oxygen is most probably correlated with the TiN phase, because the ELNES of the O*K* edge in the Ti–B–N layers corresponds closely to that of TiO [22] which is isostructural to TiN (see Fig. 5).

Recently, it has been argued that changes in orientation are the main factor in producing a variation in the ELNES in TiB_2 [21]. There may be a number of reasons for this effect. Firstly, TiB_2 has a hexagonal structure like BN, but in contrast to BN it consists of monoelement layers, *i.e.* boron layers sandwiched between layers of titanium in the *a-b* plane. It therefore has anisotropic bonding within the whole structure, as the bonding within the layers will be different to any interlayer bonding. These differences are very directional, and hence it is perhaps not surprising that orientation effects are observed. However, in the polycrystalline fine-grained sample of Ti–B–N any ELNES differences will be averaged out.

In the B*K* edge of TiB_2 the feature at *ca.* 190 eV may be assigned to transitions to π^* states, whereas the broad peak at *ca.* 202 eV is associated with σ^* states as has been recently confirmed by multiple scattering calculations [21]. It is also important to mention that the B*K* edge is only similar to TiB_2 near the edge threshold, but shows significant changes for energy losses far above the edge threshold. This is a strong indication for a disordered structure which will be discussed below.

The above findings are confirmed for the N*K* edge (Fig. 7) which closely corresponds to that of crystalline TiN, but also shows a broadening of the edge fine structures far above the edge threshold. This means that essentially there is a cubic phase of TiN nanoparticles which are uniformly distributed in the coating layer. The N*K* edges of the BN compounds differ considerably, thus confirming that there is not much boron nitride in the sample.

The comparison of the B*K* and N*K* edges of Ti–B–N with the reference compounds TiB_2 and TiN shows that the features near the edge threshold are quite similar. This means that the nearest neighbour environment in Ti–B–N is similar to the references. However, with increasing energy losses (25–30 eV above edge threshold) a broadening of the edge features is observed. These resonances have been ascribed to the scattering from the first and second neighbour [23]. Using multiple scattering calculations of rock salt structured nitrides and oxides, *Kurata et al.* have identified the broad peak with scattering from the first non-metal shell [24]. Since the B*K* edge of TiB_2 in the Ti–B–N coating (Fig. 6) is more broadened than the N*K* edge in TiN (Fig. 7), we may expect that long-range order is lost to a higher degree in case of the TiB_2 particles. This is confirmed by X-ray and electron diffraction results, where it was not possible to detect crystalline TiB_2 in the Ti–B–N coatings. These results are in agreement with previous findings [7] where EXAFS spectroscopy of Ti–B–N coatings has shown that the second coordination shell of Ti atoms was reduced to an unreasonably low coordination number (less than two atoms), indicating high average disorder beyond the first boron or nitrogen coordination shell and no real long-range order [7].

Within the Ti–B–N system, separate elemental or binary phases are thermodynamically favoured during PVD growth of the coating. It is well known that the deposition rate and surface diffusion coefficient are two important parameters in determining the grain size of deposited monophase coatings [7]. In multiphase films, the grain size is further influenced by competitive growth between grains of different composition. For the Ti–B–N multiphase system, the grain size is limited to a few nanometers when two phases are present. If three or even five

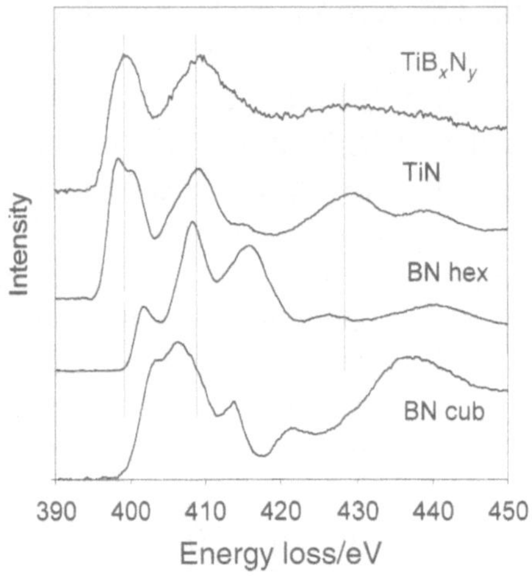

Fig. 7. Comparison of the N*K* edge of $TiB_{0.6}N_{0.7}$ from the middle position of the layer with N*K* edges from reference compounds (background below the ionization edges has been subtracted)

phases are formed, the grain size is progressively reduced to a point where well crystallized grains cannot be observed.

For a detailed information on the phases occurring in the magnetron sputtered Ti–B–N system, it is appropriate to summarize all experimental results which leads to the following model: There are two main phases TiN and TiB_2 all over the coating and at least two other phases such as B_2O_3, occurring mainly near the substrate and TiO in the middle of the layer. The well-crystallized TiN and TiO particles with an average diameter of several nanometers are embedded in a strongly disordered or *quasi*-amorphous phase consisting of TiB_2 and, near the steel substrate, also B_2O_3 (Table 2).

Finally, it has to be mentioned that techniques such as EXELFS or EXAFS have been successfully used to study bond lengths and coordination numbers in nanocrystalline materials (see Ref. [25]). However, these oscillations cannot be fully exploited, because they have to be collected from energy losses of 100 to 300 eV above the edge threshold. Since the B*K*, C*K*, or N*K* edges are only separated by *ca.* 100 eV and the N*K* edge overlaps with the Ti$L_{2,3}$ edge, the application of these techniques is very limited. Consequently, there is no choice to characterize multiphase nanocomposites by a combination of different methods such as electron and X-ray diffraction, photo electron spectroscopy, electron energy-loss spectroscopy, and near-edge fine structures techniques.

Conclusions

This experimental investigation has been concerned with understanding the microstructure and chemistry of nanocrystalline Ti–B–N layers using a combination of X-ray and electron diffraction, photoelectron spectroscopy, transmission electron microscopy, and electron energy-loss spectrometry. We can assume that the Ti–B–N nanocomposite coating consists mainly of TiN and TiO crystals with an average diameter of several nanometers which are embedded in a strongly disordered or *quasi*-amorphous phase consisting mainly of TiB_2 and, near the steel substrate, also

B_2O_3. This combined electron microscopic, XRD, and XPS study provides important information from which the physical and mechanical properties of the technological Ti–B–N materials can be understood and refined.

Experimental

Sample preparation

The unbalanced DC magnetron sputtering system has been described in detail elsewhere [12], and only essential features will be presented here. TiN and TiB_2 targets with diameters of 15 cm were cut in two halves, and one half of each was bonded onto a water-cooled backing plate. The target was positioned at a distance of 6 cm from the parallel-plate substrate assembly. The sputtering power density was maintained at approximately $2\,W \cdot cm^{-2}$. An external pair of *Helmholtz* coils was used to create a uniform axial magnetic field B_{ext}, which was adjusted between 0 and 80 G in the region between the target and the substrates to intensify the field from the outer magnetron pole. In this arrangement, coatings with a lateral gradient in the chemical composition and a typical thickness of $3\,\mu m$ were deposited. As substrates, metallographically polished austenitic stainless steel (X 5 CrNi 18 10) was used. For deposition, the ion bombardment conditions were varied systematically, (ion energy resulting from bias voltages: self-bias to $-100\,V$; ion-to-neutral flux ratio: 0.1 to 0.7).

Sample characterization

Chemical composition as well as structural and mechanical properties were investigated using electron-probe microanalysis (EPMA), scanning electron microscopy (SEM), and X-ray diffraction (XRD). The X-ray diffraction patterns were recorded with a Siemens D500 diffractometer in the *Bragg-Brentano* mode using CuK_α radiation. In addition, glancing angle XRD patterns were obtained using an unfiltered copper source at an incident angle of 1°.

XPS spectra were recorded using a VG-Escalab 210 spectrometer operating in the constant analyzer energy mode with a pass energy of 50 eV and normal detection angle. The binding energy reference was taken as the C1s peak of the carbon contamination of the samples at 284.6 eV. The X-ray source for all measurements was MgK_α (1253.6 eV). All samples were measured without surface treatment and after sputtering for 3 min with Ar^+ ions at 3 kV in the prechamber (5×10^{-6} mbar). In addition to a general spectrum over the whole energy range, zone spectra of B1s, C1s, N1s, O1s, and Ti2p with higher resolution were recorded.

TEM specimens were prepared by cross-sectioning the coatings and ion milling. The investigation was performed with a Philips CM20 TEM/STEM with an LaB_6 cathode operating at 200 kV and equipped with an energy filter (Gatan Imaging Filter, GIF) [13]. All images and EFTEM elemental maps were recorded with the slow-scan CCD camera of the GIF [14].

The spectra for the ELNES studies were acquired in the TEM image mode using a probe half angle of 1.5 mrad and a collection half angle of 7.6 mrad; a spectrometer dispersion of 0.2 eV per channel was used. The images and spectra recorded with the slow-scan CCD camera were corrected for dark current and gain variations. Although the specimens were thinner than the mean free path of inelastic scattering, the ionization edges were corrected for multiple scattering using a logarithmic *Fourier* deconvolution procedure [15]. The background below the ionization edges was subtracted using an $A \cdot E^{-r}$ model.

Acknowledgements

We gratefully acknowledge financial support by the SFB '*Elektroaktive Stoffe*' and the COST program 523 '*Nanostructured materials*'.

References

[1] Veprek S, Reiprich S (1995) Thin Solid Films **268**: 64
[2] Musil J, Vlcek J (1998) Mater Chem Phys **54**: 116
[3] Mollart TP, Baker M, Haupt J, Steiner A, Hammer P, Gissler W (1998) Surf Coat Technol **108/109**: 345
[4] Veprek S, Haussmann M, Reiprich S, Li S, Dian J (1996) Surf Coat Technol **86/87**: 394
[5] Mitterer C, Rauter M, Rödhammer P (1990) Surf Coat Technol **41**: 351
[6] Gilmore J, Baker MA, Gibson PN, Gissler W, Stoiber M, Losbichler P, Mitterer C (1998) Surf Coat Technol **108/109**: 345
[7] Baker MA, Mollart TP, Gibson PN, Gissler W (1997) J Vac Sci Technol **A15**(2): 284
[8] Mitterer C, Losbichler P, Hofer F, Warbichler P (1998) Vacuum **50**: 313
[9] Mitterer C, Mayrhofer PH, Beschliesser M, Losbichler P, Warbichler P, Hofer F, Gibson PN, Gissler W, Hruby H, Musil J, Vlcek J (1999) Surf Coat Technol **120/121**: 405
[10] Hofer F, Golob P (1987) Ultramicroscopy **21**: 379
[11] Keast VJ, Scott AJ, Brydson R, Williams DB, Bruley J (2001) J Microscopy **203**: 135
[12] Losbichler P, Mitterer C, Gibson PN, Gissler W, Hofer F, Warbichler P (1997) Surf Coat Technol **94/95**: 289
[13] Krivanek OL, Gubbens AJ, Dellby N, Meyer C (1992) Microsc Microanal Microstruct **3**: 187
[14] Hofer F, Warbichler P, Grogger W (1995) Ultramicroscopy **59**: 15
[15] Egerton RF (1996) Electron Energy-loss Spectroscopy in the Electron Microscope. Plenum Press, New York London
[16] Losbichler P, Mitterer C (1997) Surf Coat Technol **97**: 567
[17] Zanchet D, Hall BD, Ugarte D (2000) X-ray Characterization of Nanoparticles. In: Wang ZL (ed) Characterization of Nanophase Materials. Wiley-VCH, Weinheim, p 13
[18] Briggs D, Seah MP (eds) (1990) Practical Surface Analysis, 2nd edn, vol 1: Auger and X-ray Photoelectron Spectroscopy. Wiley, Chichester
[19] Schreifels JA, Maybury PC, Swartz WE (1980) J Catal **75**: 373
[20] Brydson R, Sauer H, Engel W, Zeitler E (1991) Microsc Microanal Microstruct **2**: 159
[21] Davock HJ, Tatlock GJ, Brydson R, Lawson KJ, Nicholls JR (1997) Inst Conf Ser No 153. IOP Publ Ltd, Bristol, p 609
[22] Mitterbauer C, Hofer F, Kothleitner G (2001) Proc Dreiländertagung für Elektronenmikroskopie, Innsbruck, Austria, p 32
[23] Lytle FW, Greegor RB, Panson AJ (1988) Phys Rev **B37**: 1550
[24] Kurata H, Lefevre E, Colliex C, Brydson R (1993) Phys Rev **B47**: 13763
[25] Fernández A, Sánchez-López JC, Caballero A, Martin JM, Vacher B, Ponsonnet L (1998) J Microsc **191**: 212

Received October 4, 2001. Accepted (revised) January 10, 2002

Electronic Structure and Size of TiO_2 Nanoparticles of Controlled Size Prepared by Aerosol Methods

Leonardo Soriano[1,*], **Petri P. Ahonen**[2], **Esko I. Kauppinen**[2], **Jorge Gómez-García**[1], **Carmen Morant**[1], **Francisco J. Palomares**[3], **Marta Sánchez-Agudo**[1], **Patrick R. Bressler**[4], and **José M. Sanz**[1]

[1] Departamento de Física Aplicada, Instituto de Materiales Nicolás Cabrera, Universidad Autónoma de Madrid, E-28049 Madrid, Spain

[2] VTT Chemical Technology, SF-02044 Espoo, Finland

[3] Instituto de Ciencia de Materiales de Madrid, CSIC, E-28049 Madrid, Spain

[4] BESSY, D-12489 Berlin-Adlershof, Germany

Summary. A complete characterization of nanostructures has to deal both with electronic structure and dimensions. Here we present the characterization of TiO_2 nanoparticles of controlled size prepared by aerosol methods. The electronic structure of these nanoparticles was probed by X-ray absorption spectroscopy (XAS), the particle size by atomic force microscopy (AFM). XAS spectra show that the particles crystallize in the anatase phase upon heating at 500°C, whereas further annealing at 700°C give crystallites of 70% anatase and 30% rutile phases. Raising the temperature to 900°C results in a complete transformation of the particles to rutile. AFM images reveal that the mean size of the anatase particles formed upon heating at 500°C is 30 nm, whereas for the rutile particles formed upon annealing at 900°C 90 nm were found. The results obtained by these techniques agree with XRD data.

Keywords. Electronic structure; Nanostructures; Spectroscopy.

Introduction

Nanostructured materials are of enormous interest in many technological fields due to their unique properties caused by their reduced size. However, many questions, especially those concerning the correlation of their properties with their dimensions, remain still open. To answer these questions, a characterization of the electronic structure of the nanoparticles together with an accurate determination of their dimensions is required. Usually, the electronic structure of materials is probed by electron spectroscopies. However, this is not always an easy task since, as in the case of insulating nanostructured materials, various non-controllable effects such as charging, support-nanostructure interaction, size-effects, *etc.* can affect the spectra. In this work we report on the characterization of the electronic structure of

* Corresponding author. E-mail: l.soriano@uam.es

TiO_2 nanoparticles by X-ray absorption spectroscopy (XAS). We will show below that XAS is an appropriate tool for the analysis of such nanoparticles. The fine-structure of the Ti $2p$ and O $1s$ XAS spectra allows to distinguish between the two stable TiO_2 phases, *i.e.* anatase and rutile. Moreover, XAS is able to detect and *semi*-quantitatively determine the phase composition. The particle size was determined by atomic force microscopy (AFM). The results obtained by these two surface techniques, *i.e.* TiO_2 phase structure and particle size, agree with those obtained by XRD.

In general, metal oxides are a fascinating family of materials with a large variety of electronic properties. Oxide surfaces and interfaces play an important role in many technological applications like catalysis, corrosion, sensors, *etc.* [1, 2]. In particular, titanium oxides have been widely studied due to their application as ceramics, catalysts, catalyst supports, optical coatings, gas sensors, pigments, *etc.* In all these cases, a large surface-to-volume ratio is of great importance, so that the study of TiO_2 nanoparticles seems well justified. Aerosol synthesis and, in particular, in-droplet hydrolysis of titanium alkoxide have proved to be a suitable method for the preparation of nanostructured TiO_2 particles [3, 4].

As mentioned above, the electronic structure of oxide surfaces and interfaces is usually experimentally studied by electron spectroscopies such as X-ray photoemission (XPS), *Auger* electrons (AES), electron energy loss (EELS), *etc.* [5]. The characterization of nanostructures requires techniques with adequate lateral/depth resolution. Although the lateral resolution of electron spectroscopies is progressively improving due to new technological developments, is still far from other techniques such as *e.g.* scanning tunnel microscopy (STM) that reaches atomic resolution. However, if the nanostructured material is homogeneously dispersed, electron spectroscopies can give averaged information on its electronic structure [6, 7]. XAS is an X-ray technique which probes unoccupied electronic states. In XAS, a photon is absorbed by a core electron, thus producing transitions to unoccupied states. These transitions are controlled by dipole selection rules so that only transitions to states with $\Delta l = \pm 1$ are allowed [8]. In the total electron yield detection mode, XAS is a surface sensitive technique with a mean probing depth of 40 Å [9] at the energy range used in this study (450–550 eV). This means that a near-surface region of about 20 nm depth can be probed.

Results and Discussion

Atomic force microscope images

The AFM images of the powders after annealing at 500 (a), 700 (b), and 900°C (c) are shown in Fig. 1. The images at the bottom are topographic, whereas those at the top are friction images. We have to note that previous characterization of the morphology of the powders by scanning electron microscopy (SEM) [3] had revealed spherical agglomerates (1–2 μm diameter) which showed structures in the nanometer scale. In general, the AFM images show that after pressing of the powders the initial spherical agglomerates are divided in individual nanoparticles (<200 nm) with some other bigger structures remaining. The particle size was statistically estimated for different images of the same powders. The particles

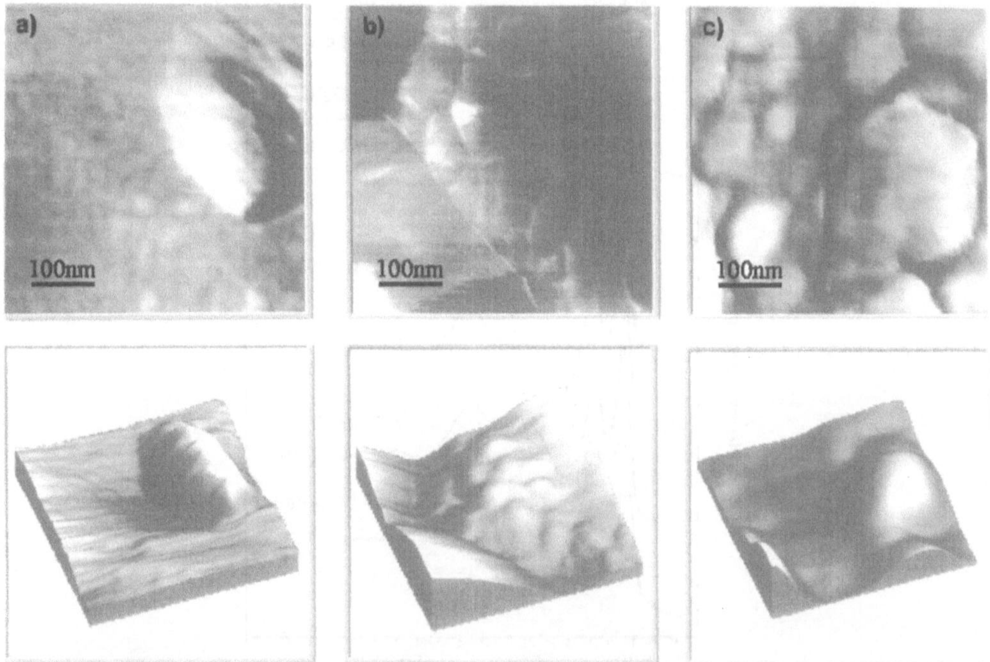

Fig. 1. AFM images of powders annealed at 500 (a), 700 (b), and 900°C (c); topographic images at the bottom, friction images at the top

Table 1. TiO_2 phase and particle size of powders annealed at different temperatures as obtained by X-ray diffraction (XRD), atomic force microscopy (AFM), and X-ray absorption (XAS)

Annealing temperature/°C	XRD [3]		AFM	XAS
	Particle size/nm	Phase	Particle size/nm	Phase
500	20	Anatase	30	Anatase
700	40–50	Anatase + Rutile	55–85	Anatase + Rutile
900	100	Rutile	90	Rutile

observed in Fig. 1a, corresponding to powders annealed at 500°C, have a mean size of 30 nm. Two average size populations were obtained for particles corresponding to powders annealed at 700°C (55 and 85 nm). For particles of powders annealed at 900°C, the mean size is 90 nm. These results are in good agreement with data obtained by XRD (Table 1). Summarizing the AFM results, we can conclude that annealing of the powders at 500°C produces very fine particles (30 nm). Posterior annealing of the powders produces an enlargement of the particle size. After annealing at 700°C, a broadening of the particle size distribution is observed, indicating that two different types of particles exist.

O 1s XAS spectra

The O 1*s* XAS spectra of the as-prepared powders and of those annealed at different temperatures are shown in Fig. 2. According to the properties of XAS

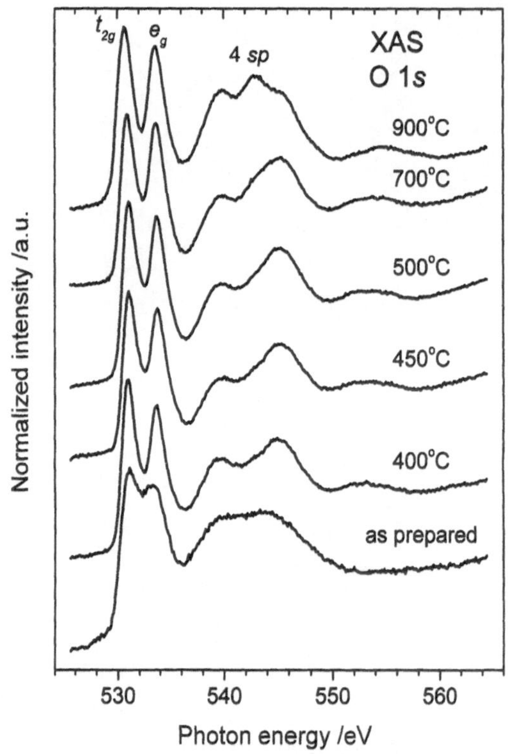

Fig. 2. O 1*s* XAS spectra of the powders as-prepared and after annealing at different temperatures

described above, these spectra correspond to unoccupied electronic states of O *p* character which are hybridized with unoccupied metal states. In fact, these spectra are mapping the metal states. In general, the spectra show two well defined sub-bands at the threshold (530–535 eV) and a broader band at higher energies (540–550 eV). The first double band originates from the hybridization of O 2*p* states with Ti 3*d* orbitals leading to the t_{2g} and e_g sub-bands. The energy splitting of these two bands, as measured in the O 1*s* XAS spectra, is very sensitive to the local structure and ligand coordination and a good estimation of the optical crystal field splitting (Δ_{CF}) [11]. The second band originates from the hybridization of O 2*p* states with Ti 4*sp* orbitals. This band appears at higher energies due to the larger overlap of the O 2*p* – Ti 4*sp* orbitals and is more sensitive to long-range order in the sample.

The spectrum of the as-prepared powders shows that the band at the threshold is broader with a lower splitting of the two sub-bands, *i.e.* a lower crystal field splitting. This is a clear indication of a weaker interaction between the Ti 3*d* and O 2*p* orbitals. According to the O 1*s* XAS spectra of Ti oxides previously reported [12] this can be interpreted in terms of a lower oxidation state of the Ti atoms corresponding to the titanium hydrous oxide as-prepared powders [3]. On the other hand, in the spectra of the annealed nanoparticles the two sub-bands at the threshold appear narrow and well defined with an energy separation of 2.6 eV, in agreement with other reported values for the crystal field splitting in TiO_2 [13]. These spectra are almost identical except for the 4*sp* band which shows three main

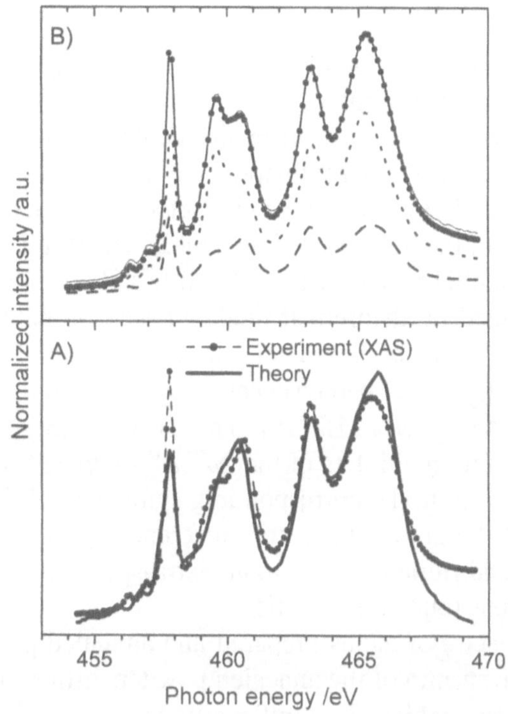

Fig. 3. A) Unoccupied density of states of O *p* character calculated in Ref. [14] for anatase (solid line) and O 1*s* XAS spectra of powders annealed at 500°C (dots); B) Unoccupied density of states of O *p* character calculated in Ref. [14] for rutile (solid line) and O 1*s* XAS spectra of powders annealed at 900°C (dots)

structures for the powders annealed at 900°C, but only two for the powders annealed at 500°C. The spectrum of the powders annealed at 700°C seems to correspond to a transition stage between those corresponding to annealings at 500 and 900°C.

To understand these differences in the 4*sp* band we have depicted the band structure calculations performed by *de Groot et al.* [14] for TiO_2 in both anatase (A) and rutile (B) phases in Fig. 3. The calculated density of states (DOS) of oxygen 2*p* character (solid line) is compared with the O 1*s* XAS spectra of the powders annealed at 500°C (A) and 900°C (B). It can be seen that the agreement of the experimental spectra with their corresponding DOS is good. The crystal field splittings of the t_{2g} and e_g bands predicted by the calculations are in complete agreement with the experimental spectra. The calculated 4*sp* bands show two main structures for the anatase phase, whereas for the rutile phase those bands show three different peaks. From the above comparison it is inferred that after annealing at 400°C anatase TiO_2 phase is formed, whereas after annealing at 900°C anatase is completely transformed to rutile. The spectrum of the powders after annealing at 700°C seems to indicate that at this temperature both phases are present. These XAS results are in complete agreement with the XRD results summarized in Table 1.

Ti 2p XAS spectra

According to the selection rules of XAS, the Ti 2*p* XAS spectra correspond to Ti $2p \rightarrow 3d$ and Ti $2p \rightarrow 4s$ transitions ($\Delta l = \pm 1$). However, the spectrum is dominated by transitions to Ti 3*d* states since the contribution of the Ti *s* states is negligible. Although in a first approximation the Ti 2*p* XAS spectra should map the unoccupied density of states of Ti 3*d* character, other final state effects affect the spectra. In particular, the hole created when a 2*p* electron is ejected to a narrow *d* band produces a potential which disturbs the original spectra; as a consequence, they cannot be explained by theoretical models based on band structure calculations. However other models based on atomic multiplets projected in the corresponding symmetry have been successfully used to explain most of the transition-metal 2*p* XAS spectra [15, 16]. The Ti 2*p* XAS spectra of TiO_2 have been interpreted as atomic multiplets of the $2p^6 3d^0 \rightarrow 2p^5 3d^1$ transitions projected in a crystal field according to the corresponding symmetry of the Ti atoms in each phase. In fact, as will be shown later, the different distortions of the octahedral symmetry of anatase and rutile give rise to an extra-splitting of the e_g bands in two sub-bands with different relative intensities.

The Ti 2*p* XAS spectra of the as-prepared and annealed powders are shown in Fig. 4. Once again, the spectra of the annealed powders differ from the spectrum of the as-prepared powders which is significantly broader and presents only four defined structures. The spectra of the annealed powders show five well defined structures (labelled from A to E) and two small peaks at the threshold. The spectra

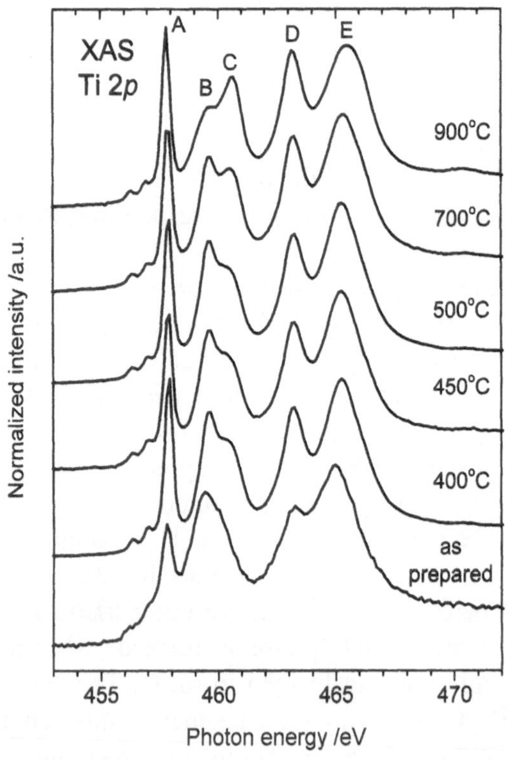

Fig. 4. Ti 2*p* XAS spectra of powders as-prepared and after annealing at different temperatures

of the powders annealed at temperatures of 400–500°C are identical and agree with other spectra published for anatase TiO_2 [10, 14], although the spectra shown in this work have a much better resolution which is a clear indication of both the good crystallinity of the nanostructured powders and a good performance of the monochromator. In these spectra, the band formed by peaks A, B and C originates from transitions from Ti $2p_{1/2}$ states, whereas the band of peaks D and E results from Ti $2p_{3/2}$ states. The different broadening of each band is related to the different lifetimes of the excited state. Peaks A and D correspond to transitions to t_{2g} states, whereas peaks B, C, and E relate to transitions to e_g states. On the other hand, the spectrum of the powders annealed at 900°C shows the same structures as the previous ones, but a different relative intensity of peaks B and C. In Fig. 5A, this spectrum is compared with atomic multiplet calculations for rutile in D_{4h} symmetry with a crystal field strength of 1.8 eV taken from Ref. [15], showing an excellent agreement. As mentioned above, the differences in the relative intensity of the B and C peaks in anatase and rutile are due to the different distorsions of the octahedrons formed in each phase [15].

According to the above XAS results we can conclude that upon annealing at 400–500°C the precursor powders crystallize to anatase TiO_2. By annealing at 900°C, anatase is completely transformed to rutile. It was demonstrated above that the Ti $2p$ spectra of anatase and rutile differ only in the relative intensity of the B

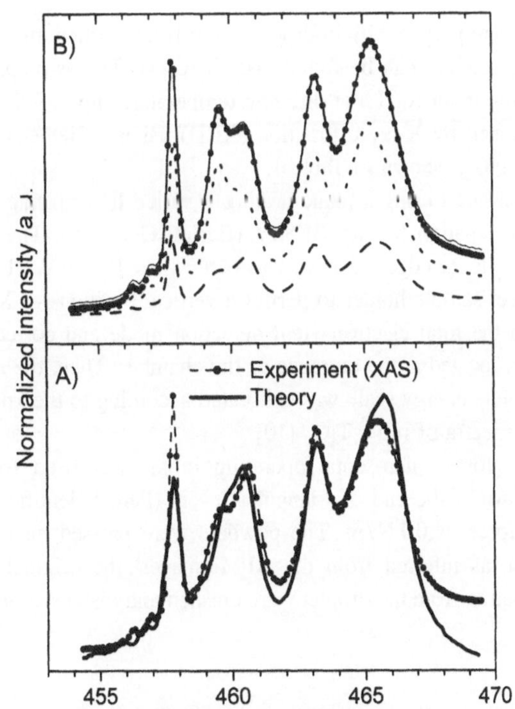

Fig. 5. A) Atomic multiplet calculations in D_{4h} symmetry for rutile with a crystal field strength of 1.8 eV taken from Ref. [15] (solid line) and Ti $2p$ XAS spectrum of powders annealed at 900°C; B) linear combination (solid line) of 30% rutile (dashed line) and 70% anatase (dotted line) compared with the Ti $2p$ XAS spectrum of powders annealed at 700°C (dots)

and C peaks. However, after annealing at 700°C the spectrum shows a relative intensity of these two bands different from those assigned to anatase and rutile. It is reasonable to think that at these temperatures anatase is only partly transformed to rutile, leading to a mixture of the two phases. In Fig. 5B the spectrum of the powders annealed at 700°C is compared with a linear combination of 30% rutile and 70% anatase. It is quite obvious that XAS is able even to quantify the relative amount of the two phases.

Conclusions

We have characterized the particle size and electronic structure of both as-prepared and annealed TiO_2 powders grown by aerosol methods. The combined AFM and XAS characterization indicates that after annealing at 500°C, particles of 30 nm size of anatase are formed. Annealing at 700°C causes 30% of the particles to be transformed to rutile with 85 nm particle size, whereas the 70% remaining anatase particles increase their size to 55 nm. Finally, annealing at 900°C produces a complete transformation to rutile particles of 90 nm size. Both XAS and AFM data are in complete agreement with XRD characterization.

Experimental

TiO_2 precursor powders were prepared by in-droplet hydrolysis of titanium alkoxide. More details on the aerosol preparation procedure can be found elsewhere [3]. The as-prepared powders were submitted to thermal annealing in air for 1 h at different temperatures up to 900°C. Crystal structure and particle size were determined by X-ray diffraction (XRD) using a Siemens D500 diffractometer. A summary of the XRD results is shown in Table 1.

XAS spectra were measured using a plane grating – varied line spacing monochromator (PGM-VLS) in the synchrotron radiation facility BESSY (Berlin, Germany). The estimated resolution of this monochromator at the Ti 2*p* edge (450 eV) was better than 100 meV. The powders were pressed on a suitable stainless steel sample holder to permit a vertical position for XAS measurements. The spectra were recorded in the total electron yield detection mode and corrected with the I_0 current measured from a gold net located at the entrance of the chamber. Then, the spectra were normalized for comparison. The absolute energy scale was calibrated according to the known position of the first peak of the Ti 2*p* EELS spectra of rutile TiO_2 [10].

A home-built atomic force microscope operating in air was used for AFM measurements. Pyramidal Si_3N_4 tips grown at the end of triangular levers (Park Scientific Instruments) were used with a normal constant force of 0.1 N/m. The powders were pressed on mica for AFM measurements. By this operation, as inferred from the AFM images, the original spherical agglomerates (1–2 μm diameter) originating from the droplet were crashed and dispersed on the surface of the mica substrate.

Acknowlegements

This work was financially supported by the EC-HPRI Programme under contract HPRI-1999-CT-0028, the MCYT of Spain, contract number BFM2000-0023, and the Comunidad de Madrid, contract number 07N-0006-1999. We thank the staff of BESSY and *J. A. Rodríguez* from UAM for technical support.

References

[1] Henrich VE, Cox PA (1994) In: The surface Science of Metal Oxides. Cambridge University Press, Cambridge

[2] Noguera C (1996) In: Physics and Chemistry at Oxide Surfaces. Cambridge University Press, Cambridge

[3] Ahonen PP, Tapper U, Kauppinen EI, Coubert JC, Deschanvres JL (2001) Mat Sci Eng A **315**: 113

[4] Ahonen PP, Richard O, Kauppinen EI (2001) Mat Res Bull **36**: 2017

[5] González-Elipe AR, Yubero F (2001) Spectroscopic Characterization of Oxide/Oxide Interfaces. In: Nalwa HS (ed) Handbook of Surfaces and Interfaces of Materials. Academic Press

[6] Sanz JM, Núñez R, Fuentes GG, Soriano L, Morant C (1999) J Surface Analysis **5**: 338

[7] Sanz JM, Soriano L, Prieto P, Tyuliev G, Morant C, Elizalde E (1998) Thin Solid Films **332**: 209

[8] Fuggle JC, Inglesfield JE (1992) In: Unoccupied Electronic States. Springer, Berlin

[9] Abbate M, Goedkoop JB, de Groot FMF, Grioni M, Fuggle JC, Hofmann S, Petersen H, Sacchi M (1992) Surf Interface Anal **18**: 65

[10] Brydson R, Sauer H, Engel W, Thomas JM, Zeitler E, Kosugi N, Kuroda H (1989) J Phys Condens Matter **1**: 797

[11] de Groot FMF, Grioni M, Fuggle JC, Ghijsen J, Sawatzky GA, Petersen H (1989) Phys Rev B **40**: 5715

[12] Lusvardi VS, Barteau MA, Chen JG, Eng J Jr, Teplyakov AV (1998) Surf Sci **397**: 237

[13] Soriano L, Abbate M, Fuggle JC, Jiménez MA, Sanz JM, Mythen C, Padmore HA (1993) Solid State Comm **87**: 699

[14] de Groot FMF, Faber J, Michiels JJM, Czyzyk MT, Abbate M, Fuggle JC (1993) Phys Rev B **48**: 2074

[15] de Groot FMF, Fuggle JC, Thole BT, Sawatzky GA (1990) Phys Rev B **41**: 928

[16] de Groot FMF, Fuggle JC, Thole BT, Sawatzky GA (1990) Phys Rev B **42**: 5459

Received October 5, 2001. Accepted (revised) December 6, 2001

Influence of Nanoporosity and Roughness on the Thickness-Dependent Coercivity of Iron Nanonetworks

Syed A. M. Tofail[1], **Zakia I. Rahman**[1,*], **Abdur M. Rahman**[2], and **Razeeb K. U. Mahmood**[1]

[1] Department of Physics and Materials and Surface Science Institute, University of Limerick, Limerick, Ireland

[2] Department of Electronic and Computer Engineering and Materials and Surface Science Institute, University of Limerick, Limerick, Ireland

Summary. We have studied the coercivity of magnetic nanonetworks as a function of thickness, nominal pore diameter, and surface/interface roughness in the thickness range of approximately 2 to 45 nm where a *Néel*-type domain wall has been theoretically predicted. Such magnetic nanonetworks have been prepared by sputtering iron on the walls of commercially available porous nanochannel alumina (NCA) membranes. The thickness dependence of coercivity has also been studied on films deposited on surface-oxidized Si and glass subtrates. These substrates are essentially non-porous and much smoother than NCAs. Our investigation shows that the coercivity of films deposited on Si and glass depends on the spatial fluctuation of thickness which arises from the roughness of the apparently smooth substrates. On the other hand, NCAs are found to be inherently quite rough, and films on NCAs show a complex thickness dependence which arises from the interplay between surface/interface roughness, domain pinning due to porosity, surface anisotropy, surface torques, and oxidation of the iron films. It was found that the growing films on NCA substrates led to partial filling up of the pore entrance, thereby reducing its effective diameter. The film growth also affects the roughness of the film, which in turn affects its coercivity. We propose a model for the thickness dependence of coercivity based on the pore fill-up geometry considering the effective pore diameter and the critical thickness at which the pore will be completely filled up. Experimental results on coercivity with thickness variation of iron network deposited on NCA generally agree with the suggested model.

Keywords. Nanostructures; Magnetic properties; Iron film; Porosity; Surface roughness.

Introduction

Fabrication of patterned structures on the nanometer scale in two or three dimensions require a suitable pattering technique such as optical or electron beam lithography, or alternatively the use of nanotemplates. Depending on the method of

* Corresponding author. E-mail: zakia.rahman@ul.ie

patterned matrix formation, either a positive or a negative nanostructured network can be obtained [1]. Many earlier works on patterned magnetic nanostructures have been focused on magnetic information storage to achieve higher recording density beyond 100 Gbits/in^2. Studying patterned media is also relevant to the understanding of fundamental magnetic phenomena. For example, *Cowburn et al.* [2] have found that investigations of the nucleation of magnetic domains and homogeneity in conventional magnetic thin films are greatly facilitated by an anti-dot array. Anisotropy barriers and the effect of domain wall pinning on hysteresis behaviour are two other examples of what can be studied from patterned nanostructures [3].

It has been reported that templated nanostructured networks of iron have shown a coercivity increase of two orders of magnitude at around 10–20 nm film thickness as compared to similar films on continuous substrates [4]. These nanonetworks were grown on the top of the walls of porous anodized alumina (commonly known as nano-channel alumina, NCA) to form a contiguous nanonetwork. The width of the pore wall varied between 15–75 nm, depending on the pore size of the alumina template. The coercivity increase is attributed to the tendency of the magnetization of the deposit near the pores to align perpendicular to the film plane, thus acting as a energy barrier for domain wall movement [5]. The whole range of the thickness dependence of the coercivity of the templated iron film is, however, far from being well understood. In order to understand the effect of porous or rough substrates it is important to gather knowledge on the film morphology originating from fabrication history and its impact on the coercivity behaviour. In this article we have attempted to study this problem from the existing theory of the thickness dependence of the coercivity of ferromagnetic materials deposited on non-porous substrates by incorporating porosity in the substrate and surface roughness in the growing film on the top of this substrate. The coercivity (H_c) of sputter-deposited ferromagnetic films is greatly influenced by sputtering conditions and magnetic layer thickness. H_c is defined as the reverse field necessary to bring residual magnetization back to zero. This is critical in determining the amount of information that can be recorded in a given area as well as the stability of these data bits with respect to long-term storage [1]; higher coercivity is desirable for such applications.

The coercivity of the ferromagnetic films depends on thickness and porosity [6], level of oxidation [7], and substrate roughness [8]. Sputtering conditions can control some of these factors. The film thickness is of primary influence on the coercivity as this factor determines both domain wall energy and domain wall type. A domain wall separates two oppositely magnetized domains and consists of a transition layer of characteristic width and energy associated with its formation and existence. In the case of an ideally smooth magnetic film it can be assumed that the magnetization of such a film should be reversed by the parallel displacement of an 180° domain wall which is parallel to the easy magnetization axis of the film so that the energy of the domain wall is a function of displacement. In Fig. 1 such a case of domain wall displacement is shown. In this case, the magnetization direction in region 1 is in the direction of applied magnetic field, and the domain wall in this region grows at the expense of the domain in region 2. For a sufficiently thick films ($D \ll t$; D: domain wall width, t: film thickness), a *Bloch*-type domain wall

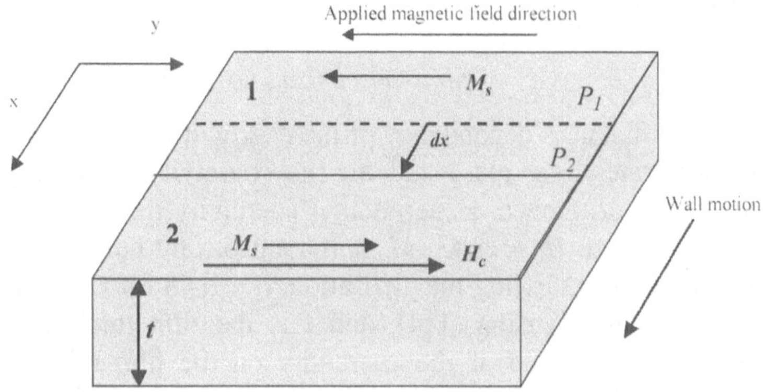

Fig. 1. Parallel displacement of a 180° domain wall from position P_1 to position P_2 in a ferromagnetic film due to an applied magnetic field [9]

can exists [9] in which the magnetization component is perpendicular to the domain wall. In this case, H_c shows the well-known $t^{-4/3}$ dependence as calculated by *Néel* [10]:

$$H_c \propto t^{-4/3} \tag{1}$$

Néel, however, has also pointed out that below a certain critical film thickness the *Bloch* wall will become energetically unfavourable, and the magnetization component will be directed parallel to the domain wall. In this situation, a *Néel*-type domain wall will from ($t \ll D$). *Sohoo* [9] has found that coercivity is independent of thickness and varies according to Eq. (2) where dt/dx is the spatial fluctuation of film thickness.

$$H_c \propto \frac{dt}{dx} \tag{2}$$

For an ideally smooth film on an ideally smooth substrate, dt/dx can be considered to be independent of the film thickness, and hence Eq. (2) should be valid for the thickness range where *Néel*-type domain walls are energetically favourable. If porosity is introduced in this hypothetical substrate, it will render the overlying film to be porous along with the fact that the pore dimensions (diameter and pore wall width) will be a function of the film thickness. This is important in determining the final coercivity of the deposit, as porosity can act as pinning sites for the domain wall, thereby restricting its motion and increasing the coercivity. Recent advances in lithography technique have allowed cutting regular holes into the ferromagnetic films and to investigate the influence of such holes on the magnetic properties of the film. As mentioned before, NCA membranes are also used, as they provide a ready-to-use template to introduce such porosity in the ferromagnetic film [4, 11]. The pores in such templates, as has been discussed elsewhere [11], are often random in geometry (size and distribution) and lack long-range order. In addition, NCA templates exhibit quite high surface roughness. Since all these parameters can influence coercivity, any study on the coercivity behaviour of ferromagnetic films grown on such templates should carefully consider the surface characteristics of the template.

Results and Discussions

Thickness dependent coercivity of films on continuous substrates

Figure 2 shows the thickness dependence of the coercivity of iron films deposited on nonporous substrates, *e.g.* glass and Si; the coercivity values of these two systems showed little difference. Experimental coercivity data followed the relation $H_c = H_0 t^{-0.75}$, where $H_0 = 2\alpha K_1/M_s$ is the anisotropy field, K_1 is the anisotropy constant, M_s is the saturation magnetisation, $\alpha = 0.48$ for random orientation of crystallites in three-dimensions [14], and t is the film thickness. A similar inverse power law dependence of the coercivity on the film thickness with an exponent value of $n = 0.4 \pm 0.1$ was found for Co films deposited on Cu buffered (111) Si by *Min et al.* [8]. The coercivity relationship along with a coercivity lower than the anisotropy field (564 Oe for iron) of the continuous films indicates that the reversal process is governed predominantly by domain wall motion rather than coherent rotation [3]. The inverse power law dependence has been explained by *Min et al.* in terms of the roughness of the ferromagnetic film [8]. For the thickness range studied, the walls were expected to be of *Néel*-type, which means that coercivity should be a function of dt/dx. The observed non-zero exponent n was believed to be due to the variation of film thickness dt/dx as a function of t. If the substrate roughness is small compared to the film thickness, then the root mean square thickness fluctuation should approximately be the sum of the interface widths of the bare substrate and the film-substrate system. Assuming a decrease in film roughness with increasing in film thickness, they suggested that the proposed decrease in the local roughness with deposition of Co could cause the coercivity to decrease by changing the magnetostatic and anisotropy energies [8].

In Fig. 3, the thickness dependence of the roughness as measured from AFM topographies and normalized to their correlation length is shown. The data were fitted using the empirical relation $\sigma/\xi = at^b$, where σ is the roughness and ξ the correlation length measured from AFM. Below 7.5 nm film thickness, the constant a changes its value from 2 to 0.009 for Si and from 3.5 to .009 for glass. The

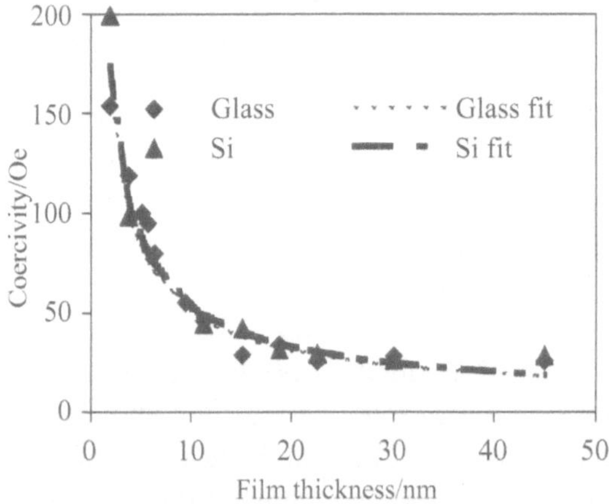

Fig. 2. Thickness dependent coercivity variation of polycrystalline iron films deposited on nonporous substrates, *e.g.* glass and Si

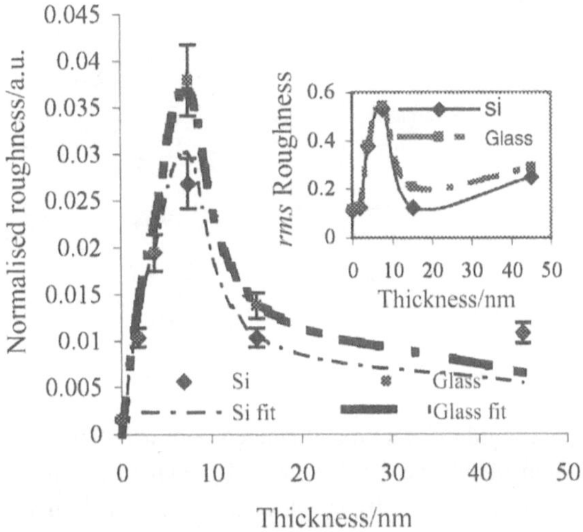

Fig. 3. Thickness variation of the normalized roughness of iron films deposited on non-porous substrates (Si and glass); inset: thickness variation of root mean square roughness of the same set of samples

exponent b changes its sign at this film thickness and assumes a value 0.6 for Si and 0.7 for glass. If we consider the film roughness variation as the spatial fluctuation of film thickness, *i.e.* $dt/dx = \sigma/\xi$, then Eq. (2) qualitatively describes the thickness dependence of coercivity in terms of the film roughness.

Thickness dependent coercivity of films on NCA templates

NCA templates are inherently quite rough; this roughness, in turn, has been found to be a function of the pore dimension of the templates [15]. The determination of the spatial fluctuation of film thickness dt/dx will therefore be complicated by the presence of porosity and, it should depend on the variation of film thickness, too. This problem can be approached in a much more convenient way if we consider the pore fill-up geometry as shown in Fig. 4. In an earlier report, we have shown the cross-sectional TEM view of a similar deposit on a porous template [11] and found that such a deposition is always associated with a certain degree of pore filling as well as pore capping. The extent to which this will happen depends on the pore diameter, the geometry of the pore opening and, most importantly, on the thickness of the overlying deposit [13]. It should be mentioned that the templates employed for this work are mainly intended for use as filtration membranes to retain oversized particles or organisms. This particular application of the templates has dictated the structure of the membranes, and the pores are tapered along the cross-section. The pore size specified by the manufacturer is only available at the filtration side of the membranes and, as evident from the cross-sectional views so far described, these pores are conical rather than cylindrical (Fig. 4). Closer observation reveals a typical trickling of the iron oxide deposit down the sidewalls of these conical pore openings. The overflow deposit down the sidewalls ends up at a finite distance below the pore entrance.

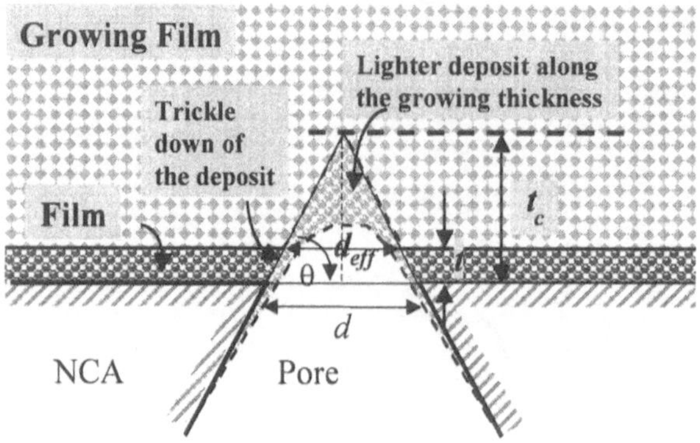

Fig. 4. Schematic diagram of filling of a typical tapered pore with nominal entrance diameter d as a result of deposition of film of thickness t; the critical thickness at which the overlying film will completely cover up the pore is denoted by t_c, whereas d_{eff} is the effective pore diameter at a given thickness and θ the angle the lateral growth of the film subtends to the pore entrance

The growth of the sputter deposited layer over the top of the magnetic layer is also important as it indicates the critical thickness where the pore will be completely filled up by the overlying deposit. A linear overgrowth of the film from the point of inflexion of the conical pores can be noticed which continues up to a certain critical thickness t_c beyond which there is no discontinuity in the overgrown layer. Figure 5 is just a redraw of Fig. 4 ignoring the trickle effect of the deposit; spatial discontinuity in the deposit due to the porous template can be noticed from this diagram. The spatial variation of thickness as well as the spatial variation of

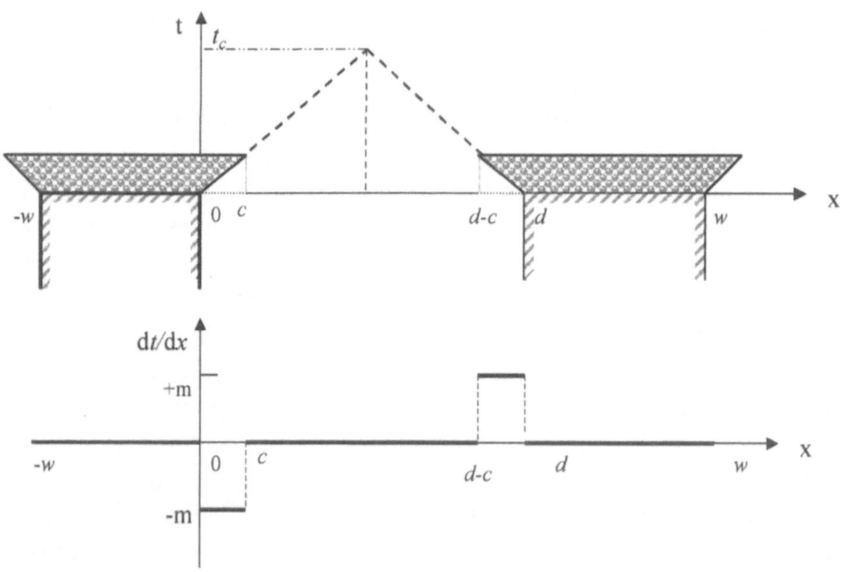

Fig. 5. Spatial variation of thickness and slope of the thickness profile (dt/dx) of an ideal smooth film on a porous substrate; the lateral extent of pore fill-up of the film is denoted by c; w: pore wall width, m: slope of the thickness profile

the slope due to thickness variation of an ideal flat film is also shown in Fig. 5. Assuming that the pores with nominal diameter d are partially filled up by an overlying film of thickness t, the pore diameter at the top of the deposit is given by Eq. (3) where c is the lateral extent of pore fill-up as described in Fig. 5.

$$d_{\text{eff}} = d - 2c \tag{3}$$

On the other hand, the critical thickness t_c for complete filling of the pores will depend on the angle θ that the deposit subtends with the nominal pore diameter. Clearly, c achieves its maximum when $\theta = 0$ and its minimum when $\theta = \pi/2$; a small change $\Delta\theta$ will change c by Δc, and t_c will be reduced by an amount of Δt_c. Then, from geometric relationships and after suitable manipulation, we get $\Delta t_c / t_c = \tan\Delta\theta / \tan\theta$. As $\Delta t_c \to t_c$ when $\Delta\theta \to 0$, $t_c = d\Delta\theta/2$. Also, $\Delta\theta$ corresponds to an equivalent change in the film thickness Δt, resulting in $\Delta\theta = \tan\theta = \Delta t/\Delta c$ and, finally, in Eq. (4).

$$t_c = \frac{d\Delta t}{2\Delta c} \tag{4}$$

This means that the critical thickness at which a conical pore in a template will be completely filled up by the overlying deposit is directly proportional to the pore diameter and the deposited layer thickness and inversely proportional to the lateral extent of pore fill-up by the deposit. For example, a film with $t = 12$ nm and $c = 8$ nm will completely cover a cylindrical pore with 200 nm nominal diameter at $t_c = 150$ nm. This can be verified from Ref. [11], where the conical pores are almost filled up by the amorphous deposit at ~150–175 nm. For a similar type of deposition, t_c can be calculated for 20 and 100 nm nominal pore diameter to be 15 and 75 nm, respectively. This means that continuous film-type behaviour should start from 150, 75, and 15 nm with nominal pore diameters of 200, 100, and 20 nm, respectively.

It can also be shown that $\Delta c \to d/2$ as $\Delta t \to t_c$, so that $c = dt/2t_c$, which means that for a given pore diameter the lateral extent of pore fill-up is a linear function of thickness. Substituting the value of c in Eq. (3) and rearranging affords Eq. (5) which gives the thickness dependence of the pore diameter of a film grown on a porous substrate.

$$d_{\text{eff}} = d\left(1 - \frac{t}{t_c}\right) \tag{5}$$

Mathematically, the thickness profile along the x-axis as can be expressed as

$$\begin{aligned} f(x) = t &= m(c - x) && \text{for} \quad 0 < x < c \\ &= 0 && \text{for} \quad c \le x < d - c \\ &= m(x - (d - c)) && \text{for} \quad d - c \le x < d \\ &= -m(d - c) && \text{for} \quad d \le x < d + w \end{aligned} \tag{6}$$

In Eq. (6), $m = 2t_c/d$ is the slope of the thickness profile along the x-axis. Similarly, the slope of the film thickness profile variation can be expressed as follows:

$$\begin{aligned} f'(x) = \mathrm{d}t/\mathrm{d}x &= -m \quad \text{for} \quad 0 \le x < c \\ &= 0 \quad \text{for} \quad c \le x < d - c \\ &= m \quad \text{for} \quad d - c \le x < d \\ &= 0 \quad \text{for} \quad d \le x < d + w \end{aligned} \tag{7}$$

If we consider that $w = d/2$, we can expand Eq. (7) into a *Fourier* series:

$$\frac{\mathrm{d}t}{\mathrm{d}x} = \sum_{n=1}^{\infty} \left(1 - (-1)^n \cos\frac{d}{2}\left(1 - \frac{t}{t_c}\right) n\pi\right) \frac{8t_c}{3n\pi d} \sin n\pi x \tag{8}$$

Taking the root mean square value over the span between $x = 0$ and $x = 3$, we obtain

$$\frac{\mathrm{d}t}{\mathrm{d}x} = 0.707 \sum_{n=1}^{\infty} \left(1 - (-1)^n \cos\frac{d}{2}\left(1 - \frac{t}{t_c}\right) n\pi\right) \frac{4t_c}{3n\pi d} \tag{9}$$

Equation (9) describes the thickness dependence of the spatial variation of film thickness for ferromagnetic films on a template with 200 nm pores. This simplified relationship becomes much more complicated with the introduction of a rough film on a porous substrate with considerable roughness. We then have to modify Eqs. (7) and (8) by considering the root mean square roughness, σ, of the film as follows:

$$\begin{aligned} f(x) &= m(c - x) + \sigma \quad \text{for} \quad 0 \le x < c \\ &= 0 \quad \text{for} \quad c \le x < d - c \\ &= m(x - (d - c)) + \sigma \quad \text{for} \quad d - c \le x < d \\ &= -m(d - c) + \sigma \quad \text{for} \quad d \le x < d + w \end{aligned} \tag{10}$$

and

$$\begin{aligned} f'(x) = \mathrm{d}t/\mathrm{d}x &= -m + \sigma' \quad \text{for} \quad 0 \le x < c \\ &= 0 \quad \text{for} \quad c \le x < d - c \\ &= m + \sigma' \quad \text{for} \quad d - c \le x < d \\ &= \sigma' \quad \text{for} \quad d \le x < d + w \end{aligned} \tag{11}$$

where $\sigma' = \mathrm{d}\sigma/\mathrm{d}x$ is the spatial fluctuation of film roughness.

We can expand Eq. (11) into a *Fourier* series as:

$$\begin{aligned} \frac{\mathrm{d}t}{\mathrm{d}x} &= \sigma'\left(\frac{2c}{3} - 1\right) + \sum_{1}^{\infty} \frac{2c\sigma'}{3n\pi} \sin n\pi c \cos n\pi x \\ &\quad - \sum_{1}^{\infty} \frac{2c\sigma'}{3n\pi} \left(2m\left(1 - (-1)^n \cos n\pi \frac{d_{\mathrm{eff}}}{2}\right) - \sigma'(1 - (-1)^n)\right) \sin n\pi x \end{aligned}$$

Since c is very small, the middle term may be ignored; averaging between $x=0$ and $x=3$ then affords

$$\frac{\mathrm{d}t}{\mathrm{d}x}=\sigma'\left(\frac{2c}{3}-1\right)-0.707\sum_{1}^{\infty}\frac{2c\sigma'}{3n\pi}\left(\frac{2t_c}{d}\left(1-(-1)^n\cos n\pi\frac{d}{2}\left(1-\frac{t}{t_c}\right)\right)\right.$$
$$\left.-\sigma'(1-(-1)^n)\right) \qquad (12)$$

We can then modify Eq. (2) for the *Néel* wall zone as

$$H_c\propto\sigma'\left(\frac{2c}{3}-1\right)-0.707\sum_{1}^{\infty}\frac{2}{3n\pi}\left(\frac{2t_c}{d}\left(1-(-1)^n\cos n\pi\frac{d}{2}\left(1-\frac{t}{t_c}\right)\right)\right.$$
$$\left.-\sigma'(1-(-1)^n)\right)$$

or

$$(H_c)_{\text{roughness}}\propto\sigma'\left(\frac{2t}{3t_c}(d-1)\right)-0.707$$
$$\sum_{1}^{\infty}\frac{2}{3n\pi}\left(\frac{2t_c}{d}\left(1-(-1)^n\cos n\pi\frac{d}{2}\left(1-\frac{t}{t_c}\right)\right)-\sigma'(1-(-1)^n)\right) \qquad (13)$$

For 20 nm pores, $d\approx w$ rather than $2w$ [4]; Eq. (12) can therefore be modified according to Eq. (14)

$$\frac{\mathrm{d}t}{\mathrm{d}x}=\sigma'\left(\frac{t}{2t_c}(d-1)\right)-0.707\sum_{1}^{\infty}\frac{4t_c}{n\pi d}\cos\frac{\mathrm{d}}{2}\left(1-\frac{t}{t_c}\right)n\pi \qquad (14)$$

which finally affords

$$(H_c)_{\text{roughness}}\propto\sigma'\left(\frac{t}{2t_c}(d-1)\right)-0.707\sum_{1}^{\infty}\frac{4t_c}{n\pi d}\cos\frac{d}{2}\left(1-\frac{t}{t_c}\right)n\pi \qquad (15)$$

In addition to roughness there will be two other opposing factors working on the magnetization reversal in such systems: domain wall pinning and domain wall nucleation. Whereas porosity in the pinning model hinders domain wall motion and thus increases coercivity, it can also act as a nucleation centre for the nucleation of a domain wall and thereby decrease coercivity.

Hilzinger and *Kronmüller* [16] have considered the pinning of the curved domain walls by randomly distributed defects, and the coercivity was found to depend on $\rho^{1/2}$ for weak pinning and on $\rho^{2/3}$ for strong pinning (ρ: density of inclusions/voids/pores). More recent theories suggest the following conditions for domain wall pinning: $3f/2\pi\gamma\delta<1$ for weak pinning and $3f/2\pi\gamma\delta>1$ for strong pinning (f: maximum restoring force a pinning centre can exert on a domain wall, γ: domain energy per unit area, δ: domain wall thickness) [17, 18].

The restoring force can be calculated by using the *Néel* model for unpaired spins at the surface of a nonmagnetic precipitate [19]. If we assume cylindrical pores, the restoring force then can be calculated as $f=0.5(M_s^2N_d)(V_p/r)$ where V_p

is the volume of the pore, N_d is the demagnetizing field associated with a cylindrical pore ($\approx 1/3$), and r is the radius of the pore. From the volume of the frustum in Fig. 4 it can be concluded that the pore volume at a given thickness t of the film is

$$V_p = \frac{\pi}{12} d^2 \left(1 - \left(1 - \frac{t}{t_c}\right)^3\right) t_c \tag{16}$$

Hence,

$$f = \frac{\pi}{36} M_s^2 \left(1 - \left(1 - \frac{t}{t_c}\right)^3\right) dt_c \tag{17}$$

Taking $\gamma \approx 2\,\mathrm{erg/cm^2}$ and $\delta \approx 6\times10^{-5}\,\mathrm{cm}$ for bulk iron, we can rewrite the condition for strong or weak pinning as $2 \times 10^9(1-(1-t/t_c)^3)dt_c > 1$ for strong pinning and $2 \times 10^9(1-(1-t/t_c)^3)dt_c < 1$ for weak pinning. Figure 6 shows the variation of pinning strength with thickness for films deposited on templates with three different nominal pore diameters. Clearly these pore diameters fall within the range where only weak pinning on the domain wall movement is to be expected.

For the coercivity field in the case of weak domain pinning, *Gaunt* [17] suggested

$$H_c = \frac{0.258 f^2 \rho}{M_s \gamma} \tag{18}$$

The number of voids which are intersected by a single domain wall is roughly estimated to be $N^{2/3}$ [20], so that there are $\rho = N^{2/3}/D'$ voids in a unit volume which leads to a decrease in the magnetostatic energy (D': equilibrium separation of domain walls). If a domain wall bisects a pore of diameter d, the free pole areas are divided into regions of opposite polarity [20], thus reducing the magnetostatic

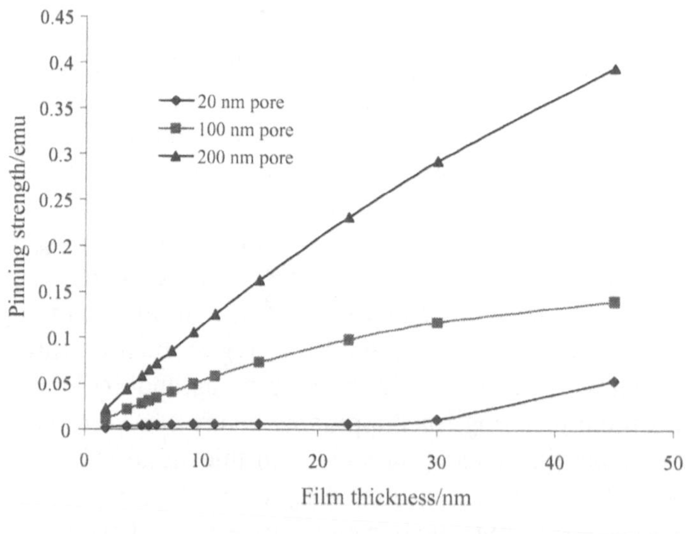

Fig. 6. Variation of domain wall pinning strength due to porosity as a function of thickness

energy to about half the value of that associated with the void. The decrease in total magnetostatic energy can then be found from the condition $N^{2/3} \cdot \Delta E_{ms} \geq \gamma$, where ΔE_{ms} is the decrease in magnetostatic energy due to the presence of one pore ($\Delta E_{ms} = 1/4 M_s^2 N_d V_p$). The equilibrium spacing of the domain wall is given as $D' = (\gamma t / C M_s^2)^{1/2}$ where t is the film thickness and C is a numerical constant that depends on the type of domain. Hence, we can calculate the number of the voids/pores/inclusions per unit volume as

$$\rho = \frac{144 C^{1/2} \gamma^{1/2}}{\pi M_s} \cdot \frac{t^{-1/2}}{d^2 \left(1 - \left(1 - \frac{t}{t_c}\right)^3\right) t_c} \tag{19}$$

Combining Eqs. (17), (18), and (19) we obtain

$$H_c = \frac{0.258 C^{1/2} M_s^2}{9 \gamma^{1/2}} \left(1 - \left(1 - \frac{t}{t_c}\right)^3\right) t_c t^{-1/2}$$

or

$$(H_c)_{\text{pinning}} \propto \left(1 - \left(1 - \frac{t}{t_c}\right)^3\right) t_c t^{-1/2} \tag{20}$$

This is shown in Fig. 7 where the normalized coercivity (H_c) is plotted as a function of film thickness for three different pore sizes.

When considering the coercivity of the iron film, we have to take into account all factors at the same time. Combining contributions of film roughness and domain wall pinning due to porosity on the top of a nonporous substrate, we obtain the qualitative relationship

$$H_c \propto (H_c)_{\text{nonporous}} (H_c)_{\text{roughness}} (H_c)_{\text{pinning}} \tag{21}$$

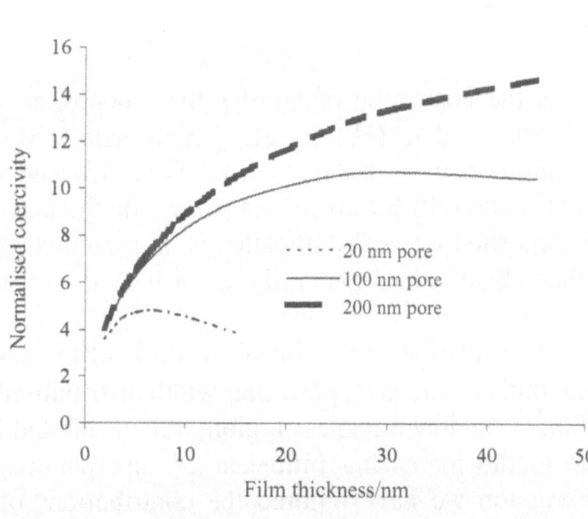

Fig. 7. Normalized coercivity as a function of thickness for templates with three different pore sizes according to the domain wall pinning model

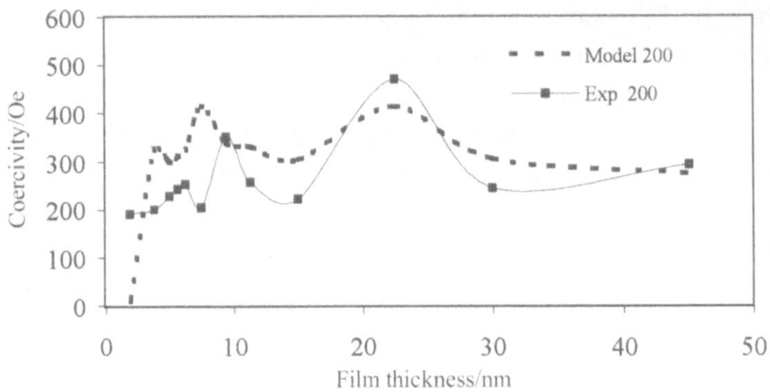

Fig. 8. Thickness dependent coercivity of iron films on NCA templates with 200 nm nominal pore diameters; broken lines: pore fill-up model, dotted lines: experimental values

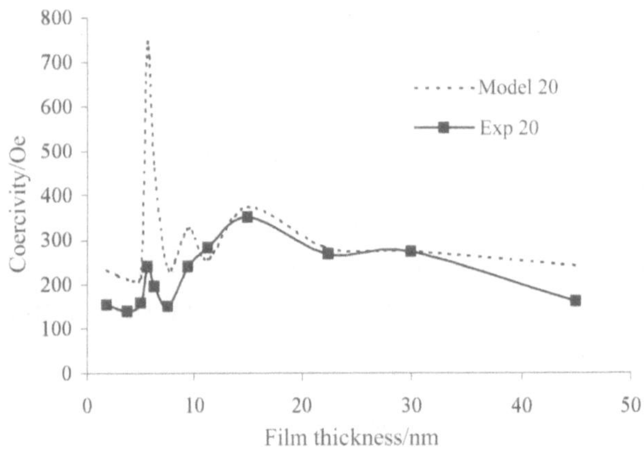

Fig. 9. Thickness dependent coercivity of iron films on NCA templates with 20 nm nominal pore diameters; broken lines: pore fill-up model, dotted lines: experimental values

where $(H_c)_{nonporous}$ is the coercivity of similar films deposited on a nonporous substrate. Figure 2 shows that $(H_c)_{nonporous}$ varies with $t^{-0.701}$. Taking this into account and using coercivity relations from Eqs. (13) and (20) for 200 nm templates and Eqs. (15) and (20) for 20 nm templates, the thickness dependence of coercivity of films deposited on such templates is as presented in Figs. 8 and 9, respectively. The theoretical values generally agree with the experimental values (solid curves).

An exact quantitative relationship is, however, difficult to establish due to the statistical nature of template pore size, pore wall width distribution, and the surface roughness of the film. In the low thickness region, films exhibited lower coercivity than predicted due to the increasing influence of superparamagnetic particles. Throughout the discussion we have ignored the contributions of surface torque and anisotropy energy as well as that of oxidation. These can influence the coercivity, especially in the low thickness region.

Conclusions

We have studied the effect of porosity and roughness on the thickness dependence of the coercivity of thin iron films in the thickness range where *Néel*-type domain wall can exist. Porous alumina templates were used to introduce porosity in the overlying film. We have developed a model on the basis of the obstruction to the domain wall motion due to the presence of pores and surface roughness in the iron film and compared this with both theoretical and experimental observations for iron films deposited on ideal pore-free smooth substrates. The model generally explains the experimental observations on the thickness-dependent coercivity of iron films deposited on nanochannel alumina substrates.

Experimental

Nanonetworks of polycrystalline iron were prepared by RF-diode sputtering of a 6″ iron target (99.95% purity) at 4 mtorr Ar pressure and 400 W sputtering power on commercially available NCA membranes [12] as substrates. The target-to-substrate distance for sputtering was kept constant at 15 cm and the base pressure below 2×10^{-7} mbar. NCA membrane templates are available with three nominal pore diameter sizes (20, 100, and 200 nm); these templates were used for deposition of iron films. Figure 10 shows a typical atomic force microscope (AFM) topography of such a template with 20 nm nominal pore diameter. The pores were found to exhibit a *Gaussian* distribution [13]. When any film is deposited on the pore walls of such templates, it produces a contiguous nanonetwork which can be considered as a negative or anti-dot nanostructured matrix. The presence of an underlying porous template introduces porosity in the overlying network of deposited films. Using similar deposition parameters, iron films were deposited on substrates that do not have any porosity such as surface oxidized (100) Si and glass substrates.

The thickness of the films was measured from cross-sectional TEM on continuous films on Si; surface roughness was determined using a Topometrix™ atomic force microscope (AFM) with an Explorer™ head in the non-contact mode. Hysteresis loops of the samples were determined using a vibrating sample magnetometer (VSM) with a maximum applied field of 3 kOe. The magnetic field was applied along the macroscopic film plane. Coercivity was measured as a function of layer thickness (2–45 nm) from the respective hysteresis loops.

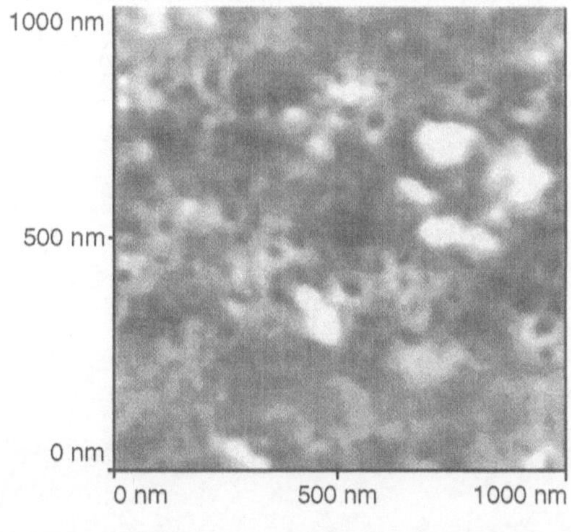

Fig. 10. Typical AFM topography of an NCA membrane with 20 nm nominal pore diameter

Acknowledgements

This project is funded by the Higher Education Authority (HEA), Ireland.

References

[1] Tofail SAM, Rahman IZ, Rahman MA (2001) Appl Organomet Chemistry **15**: 373
[2] Cowburn RP, Adeyeye AO, Bland JAC (1997) Appl Phys Letter **70**: 2309
[3] Tofail SAM, Rahman IZ, Rahman MA (2001) Thickness and Pore Size Dependence of Coercivity for Nanonetwork of Iron Produced by Template Synthesis. Presented at the MMM 2001, Seattle, Washington, Nov 12–16, 2001, to be published in J Appl Phys 2002
[4] Barnard JA, Fujiwara N, Inturi VR, Jarratt JD, Scharf TW, Weston JL (1996) Appl Phys Lett **69**: 2758
[5] Butera A, Weston JL, Barnard JA (1997) J Appl Phys **81**: 7432
[6] Hong M, Gyorgi EM, van Dover RB, Nakahara S, Bacon DD, Gallagher PK (1986) J Appl Phys **59**: 551
[7] Kim YK, Oliviera M (1993) J Appl Phys **74**: 1233
[8] Min H-G, Kim S-H, Li M, Wedding JB, Wang G-C (1998) Surface Science **400**: 19
[9] Sohoo RF (1965) Magnetic Thin Films. Harper & Row, New York, p 142
[10] Néel L (1956) J Phys Rad **17**: 250
[11] Tofail SAM, Rahman IZ, Rahman MA, Newcomb S, Sutton D (2002) J Magn Magn Mat (in press)
[12] Anopore^{TM} Whatman Inc.
[13] Tofail SAM (2002) Patterned Magnetic Nanonetworks on Nanochannel Alumina Templates. PhD Thesis, University of Limerick
[14] Sellmyer DJ, Shan ZS (1997) Magnetic Hysteresis in Novel Nanostructured Films. In: Hadjipanayis GC (ed) Magnetic Hysteresis in Novel Magnetic Materials. Kluwer, Netherlands, pp 419–451
[15] Tofail SAM, Rahman IZ, Rahman MA (2001) Templating for the Fabrication of Nanostructured Network of Iron, Proc 1st IEEE Conference on Nanotechnology (IEEE-NANO 2001), Hawaii, USA, Oct 28–31, 2001, pp 223–228
[16] Hilzinger HR, Kronmüller H (1977) Physica **86-8B**: 1365
[17] Gaunt P (1983) Philos Mag B **48**: 261
[18] Gaunt P (1986) J Appl Phys **59**: 4129
[19] Néel L (1944) Cahiers de Physique **4**: 21
[20] Chikazumi S (1997) Physics of Ferromagnetism. Oxford, p 463

Received October 16, 2001. Accepted (revised) January 11, 2002

Influence of Al_2O_3 Nanoparticles on the Thermal Stability of Ultra-Fine Grained Copper Prepared by High Pressure Torsion

Jakub Čížek[1,a,*], **Ivan Procházka**[1], **Radomír Kužel**[1], and **Rinat K. Islamgaliev**[2]

[1] Faculty of Mathematics and Physics, Charles University, CZ-18000 Prague, Czech Republic

[2] Institute of Physics of Advanced Materials, Ufa State Aviation Technical University, Ufa 450000, Russia

Summary. Ultra-fine grained (UFG) Cu (grain size 80 nm) containing 0.5 wt.% Al_2O_3 nanoparticles (size 20 nm) was prepared by high pressure torsion (HPT). Positron lifetime spectroscopy was employed to characterize the microstructure of this material, especially with respect to types and concentration of lattice defects. The evolution of microstructure with increasing temperature was studied by positron lifetime spectroscopy and X-ray diffraction measurements. The thermal stability of the Cu + 0.5 wt.% Al_2O_3 nanocomposite was compared with that of pure UFG Cu prepared by the same technique. The processes taking place during thermal recovery of the initial nanoscale structure in both studied materials are described.

Keywords. Nanostructures; Spectroscopy; Positron annihilation.

Introduction

Recently it has been found that ultra-fine grained (UFG) materials can be produced using severe plastic deformation. Grain sizes from 50 to 150 nm in metals were achieved by high pressure torsion (HPT) [1, 2]. The equal channel pressing technique (ECAP) [1] represents another method based on the above phenomenon. The UFG materials prepared by ECAP exhibit grain sizes of about 200 nm [3]. However, more massive samples can be prepared by ECAP, whereas the size of the samples prepared by HPT is limited [3]. Smaller grain sizes can be produced by HPT; therefore, this method was choosen for the preparation of the samples in this work. In comparison with other techniques of preparation of nanocrystalline

* Corresponding author. E-mail: jcizek@mbox.troja.mff.cuni.cz

[a] Current address: Institut für Materialphysik, Universität Göttingen, D-37073 Göttingen, Germany

materials (*e.g.* gas condensation or ball milling), HPT and ECAP are capable of producing high-purity specimens with no residual porosity.

TEM investigation of UFG Cu and Ni prepared by HPT [4, 5] have revealed a fragmented structure with high-angle misorientation of neighboring grains. Dislocations are concentrated in distorted layers along grain boundaries (GBs), whereas the grains themselves are almost free of dislocations. A highly non-uniform distribution of dislocations was also confirmed by XRD [6]. Most of the GBs are in non-equilibrium states of higher energy. The specific structure of UFG materials leads to a number of unusual physical and mechanical properties. In particular, the UFG materials exhibit abnormally high diffusion activity [7], unusual changes in *Curie* temperature, saturation magnetization, and elastic properties [8]. Moreover, the UFG materials are characterized by combination of high strength and ductility [1, 9]. These advantageous mechanical properties make them highly attractive for further industrial applications.

Valuable information about defects in UFG materials has been obtained by means of positron lifetime (PL) spectroscopy [5, 10]. Two types of defects were found in UFG Cu: (*i*) dislocations inside the distorted regions near to GBs and (*ii*) microvoids consisting of a few vacancies distributed homogeneously throughout the grains.

The evolution of UFG structures with increasing temperature is interesting from a physical point of view as one obtains information about the thermal recovery of the highly non-equilibrium initial structure. Moreover, detailed knowledge about thermal stability is necessary for further industrial applications of these materials. It has been found that thermal recovery of UFG structures is realized by abnormal grain growth followed by recrystallization, *i.e.* grain growth in the whole volume of the sample [5, 11]. Recently, it has been proposed that ceramic nanoparticles may inhibit the grain growth and, thereby, extend the thermal stability of the UFG structure. Indeed, investigations of UFG Cu with 0.5 wt.% Al_2O_3 have revealed enhanced thermal stability of this nanocomposite compared to pure UFG Cu [12, 13]. The onset of the recrystallization was shifted from 170°C in pure UFG Cu [4] to 400°C in UFG Cu + 0.5 wt.% Al_2O_3 [12]. However, relaxation of elastic strain and some recovery of defects indicated by a decrease of the electrical resistance precede the recrystallization [12]. Changes of microstructure corresponding to this recovery remain still unclear, but it is believed that they are connected with some changes of defect structure. PL spectroscopy represents an important tool for the investigation of these changes due to its high sensitivity to open volume defects [14].

In the present work, the thermal evolution of the microstructure of UFG Cu + 0.5 wt.% Al_2O_3 prepared by HPT was studied by PL spectroscopy and XRD, the main attention being focused on changes of the defect structure. The obtained results were compared with those for pure UFG Cu prepared by the same technique.

Results and Discussion

As-prepared state

The PL spectrum of pure UFG Cu can be fitted best by three exponential components; the lifetimes and relative intensities of these components are shown

Table 1. Positron lifetimes and corresponding relative intensities ($I_1 + I_2 + I_3 = 100\%$) for as-prepared UFG Cu and UFG Cu + 0.5 wt.% Al_2O_3; the values in parentheses represent the measurement errors (one standard deviation)

Specimen	τ_1 (ps)	I_1 (%)	τ_2 (ps)	I_2 (%)	τ_3 (ps)	I_3 (%)
Pure UFG Cu	67(8)	3.3(5)	166(2)	72(2)	255(2)	25(1)
UFG Cu + 0.5 wt.% Al_2O_3	–	–	161(3)	60.4(5)	257(1)	39.6(5)

in Table 1. The shortest component's lifetime τ_1 is well below 100 ps; this is remarkably lower than the lifetime of positrons in defect-free Cu which amounts to 114.5 ± 0.1 ps [10]. It is well known that when open volume defects are present, the lifetime of free positrons is shortened due to positron trapping at the defects [14]. The lifetimes of positrons trapped at open volume defects are always higher than the lifetimes of free positrons in defect-free material. Thus, the component of lifetime τ_1 can be obviously attributed to free positrons. The small relative intensity of this component clearly indicates that the majority of positrons annihilate from trapped states at defects. The lifetime τ_2 of the second component corresponds to the lifetime of 164 ps of positrons trapped at dislocations [15, 16]. This component originates from positrons trapped at dislocations in distorted regions along GBs; for a detailed discussion, see Refs. [5, 10]. The third component (lifetime τ_3) represents a contribution of positrons trapped at microvoids inside grains [5, 10]. Using theoretical calculations in Ref. [5], one obtains a size of the microvoids in the as-prepared specimen corresponding to 5 monovacancies, meaning that the diameter of the microvoids is about 0.5 nm. Some rare microvoids were found by TEM inside grains in pure UFG Cu [5]. The formation of microvoids of similar size was observed by PL spectroscopy also on plastically deformed polycrystalline Cu [16]. It is reasonable to assume that there is some size distribution of the microvoids, and only the largest microvoids are visible by TEM. In principle, there is no reason why the microvoids cannot be situated also inside the distorted regions. To solve this question we used the diffusion model of positron behaviour in UFG materials as described in Ref. [5]. A detailed description of the diffusion model is out of the scope of the present work; therefore, we give here only the basic ideas of the model and we will discuss the results of its application. We assume that a UFG or a nanocrystalline material consists of spherical non-distorted regions (grain interiors) with radius R which are separated by distorted regions of thickness δ. The dislocation density inside the non-distorted regions is below the detection limit of PL spectroscopy (about $10^{12}\,m^{-2}$). On the other hand, the distorted regions are characterized by high dislocation density. Thus, we assume that positrons annihilate from trapped states at dislocations inside the distorted regions only. Contrary to coarse-grained materials, the volume fraction of the distorted regions cannot be neglected in the UFG and the nanocrystalline materials. Therefore, the positrons are divided into two groups:

1) positrons which end their thermalization inside the distorted regions (we assume that they are trapped at dislocations there) and

2) positrons which end their thermalization inside the non-distorted regions; these may a) annihilate from the free state, b) be trapped in microvoids, or c) diffuse to distorted regions and be trapped at a dislocation there.

To solve this problem one has to solve the positron diffusion equation which was done similar to *Dupasquier et al.* [17]. As a result we obtained density of free positrons $n(\mathbf{r}, t)$ at a given time t and at position $\mathbf{r}$. From the density of the free positrons one can calculate the densities of the positrons trapped at the defects and deduce the shape of a PL spectrum which can be directly fitted to the shape of the experimental one. As a consequence, we can determine the linear size of the non-distorted regions, $2R$, which corresponds to the coherent domain size [5], the volume fraction of the distorted regions, η, and the density of the dislocations and concentration of the microvoids.

Regarding the microvoids, we used two different approaches in the diffusion model: (*i*) we assumed that the microvoids are distributed homogeneously inside grains only, *i.e.* they are not situated inside the distorted regions, and (*ii*) in an alternative approach the microvoids were assumed to be distributed homogeneously throughout the whole material, *i.e.* also inside the distorted regions. Approach (*ii*) leads to unphysically small grain sizes (below 10 nm), whereas approach (*i*) gave the grain sizes plotted in Fig. 9 and listed in Table 2 and which agree well with the coherent domain sizes determined by XRD. Thus, we conclude that the assumption that the microvoids are situated inside grains only is much more realistic, and approach (*i*) was used for elaboration of PL data in this work. For a detailed description of both approaches, see Ref. [5].

Two components were resolved in UFG Cu + 0.5 wt.% Al_2O_3 (see Table 1). Similarly to pure UFG Cu, the component with lifetime $\tau_2 = 161$ ps results from positrons trapped at dislocations inside the distorted regions, whereas the component with $\tau_3 = 257$ ps represents a contribution of positrons trapped at the microvoids inside grains. No free positron component was found in UFG Cu + 0.5 wt.% Al_2O_3. This may be a consequence of smaller grain size, as the relative intensity of this component is very small even in pure UFG Cu.

Whereas the lifetimes of trapped positrons are virtually the same for both specimens, the relative intensity of the component originating from the microvoids is remarkably higher for UFG Cu + 0.5 wt.% Al_2O_3. Thus, the size of the microvoids is the same in both specimens, but their concentration is higher in UFG Cu + 0.5 wt.% Al_2O_3.

Table 2. The coherent domain size a obtained by XRD and the size of the non-distorted regions $2R$ calculated from PL spectra using the diffusion model [5]; for comparison, the mean grain size d determined by TEM [4, 12] is included; the values in parentheses represent the measurement errors (one standard deviation)

Specimen	a (nm)	$2R$ (nm)	d (nm)
Pure UFG Cu	80(10)	80(4)	107
UFG Cu + 0.5 wt.% Al_2O_3	50(10)	64(4)	80

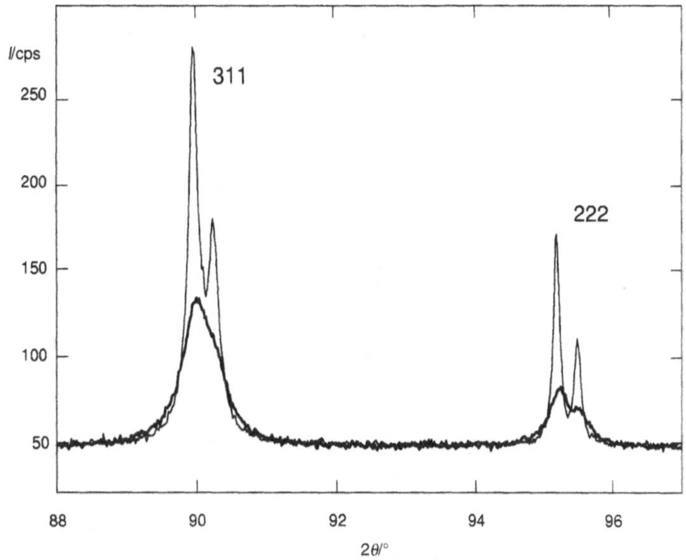

Fig. 1. XRD profiles (311) and (222) for UFG Cu with 0.5 wt.% Al_2O_3 after preparation (thick line) and after annealing at 550°C (thin line)

The lattice parameter determined for the pure UFG specimen by XRD measurements (3.6152 Å) agrees well with the values given for PDF-2 [18] (3.6150 Å). This means the absence of both impurities in the lattice and residual stresses. Texture indices show a random grain distribution or a weak (111) texture which does not vary with annealing temperature. The texture does not change during the recrystallization.

A typical feature of the XRD spectra for both UFG specimens is a broadening of the diffraction profiles which decreases with annealing temperature. This is illustrated in Fig. 1, where the measured XRD peaks (311) and (222) for as-prepared UFG Cu + 0.5 wt.% Al_2O_3 and for the specimen after annealing at 580°C are plotted. A significant strain anisotropy of line broadening of the type $\beta_{hhh} < \beta_{h00}$ for the as-deformed state was found in both samples. This anisotropy seems to be a typical feature of these materials and was found also for milled Cu powder [13]. *Williamson-Hall* (*WH*) plots for as-prepared UFG Cu and UFG Cu + 0.5 wt.% Al_2O_3 are shown in Fig. 2. The anisotropy can be related to the dislocation-induced line broadening, and the integral breadths can be calculated using known orientation (contrast) factors for the most common slip systems in *fcc* structures (*Burgers* vector $\mathbf{b} = a/2\langle 110\rangle$). A simplified procedure for the evaluation of β is similar to that published by *Ungar et al.* [19] and it is based on Eq. (1) where D is the mean domain size in the measured direction, λ is the wavelength, Θ is the diffraction angle, k is the *Scherrer* constant, b_0 is related to stacking-fault probability, $W(g)$ is the known corresponding orientation factor, $B_0 = A_0 b\sqrt{2\ln P}$, ϱ is the dislocation density, b is the magnitude of *Burgers* vector, $A_0 \sim 1$, P is a factor related to the cut-off radius of the dislocation arrangement which must be estimated from the

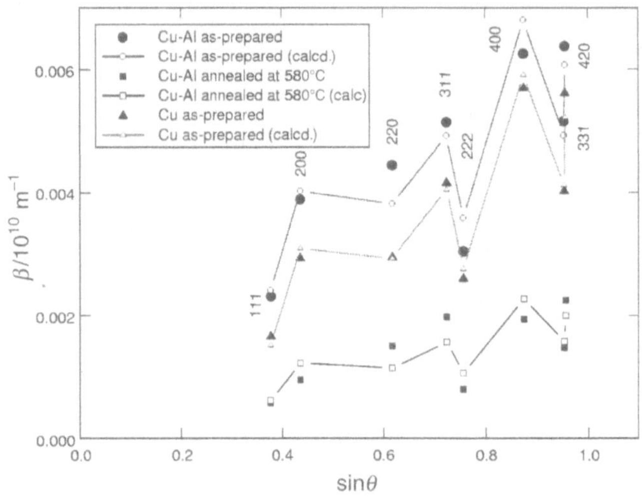

Fig. 2. *WH* plot for as-prepared pure UFG Cu (full triangles) and UFG +0.5 wt.% Al_2O_3 in the as-prepared state (full circles) and after annealing at 580°C (full squares); calculated integral breadths are indicated by open symbols connected by solid lines

diffraction profile shape (*e.g.* from *Fourier* coefficients), and O^2 indicates non-interpreted high-order terms.

$$\beta = \frac{k}{D} + b_0 W(g) + B_0 \sqrt{\varrho} \sqrt{\langle \kappa \rangle} \frac{\sin\Theta}{\lambda} + O^2 \tag{1}$$

For a mean value of the orientation factor, the *Ungar* and *Tichy* formula (Eq. (2)) is used in such a way that the fraction of edge dislocations is weighted by w_e (Eq. (3)).

$$\langle \kappa \rangle = A + BH^2 \quad H^2 = \frac{h^2k^2 + h^2l^2 + l^2k^2}{(h^2 + k^2 + l^2)^2} \tag{2}$$

$$\langle \kappa \rangle = w_e(A_e + B_e H^2) + (1 - w_e)(A_s + B_s H^2) \tag{3}$$

The parameters A_e, B_e, A_s, and B_s for the edge and screw dislocations, respectively, can be found in Ref. [20]. A two-step procedure [13] was used to determine D, b_0, w_e, and ϱ under the assumption that the *hkl*-dependence of D and the higher-order terms O^2 can be neglected. This simplified procedure gives reasonable results, especially for the dislocation densities which are the dominant reason of line-broadening. For a more detailed analysis, a new procedure should be applied [21]. The calculated values are also shown in Fig. 2 by open symbols connected by solid lines. The agreement of experimental results and theoretical calculations is excellent. The coherent domain size a, which can be obtained from *WH* plots [22], is shown in Table 2 for both the specimens. The dislocation densities for both materials as determined by XRD are plotted in Fig. 3.

Using the diffusion model of positron behaviour in UFG materials (described in details in Ref. [5]) it is possible to calculate the linear size $2R$ of the non-distorted regions, *i.e.* the regions free of dislocations. The diffusion model can be modified

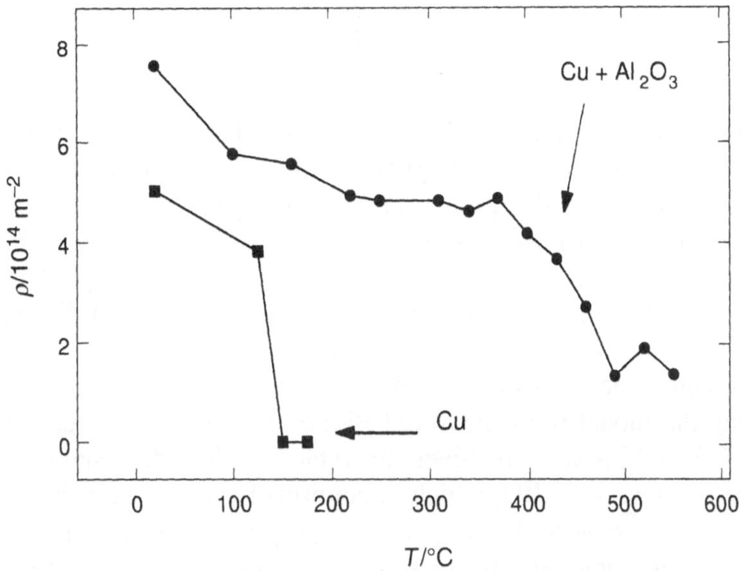

Fig. 3. Temperature dependence of the mean dislocation density determined by XRD for UFG Cu (squares) and UFG Cu + 0.5 wt.% Al_2O_3 (circles)

also to the case of saturated trapping at dislocations and microvoids which occurs in UFG Cu + 0.5 wt.% Al_2O_3. A detailed description of this modification is out of the scope of this paper and will be published separately. When applying the diffusion model to the case of saturated positron trapping in UFG Cu + 0.5 wt.% Al_2O_3, we used the additional assumption that the number of dislocations inside the distorted layers in as-prepared UFG Cu + 0.5 wt.% Al_2O_3 is the same as in pure UFG Cu. This seems to be reasonable due to the same procedure of preparation. Moreover, we can make a very simple estimation of dislocation density inside the distorted regions from the mean dislocation densities determined by XRD (Fig. 3). It is clear from Fig. 3 that the mean dislocation density in UFG Cu with Al_2O_3 exceeds that in pure UFG Cu by a factor of about 1.5. This is clearly due to the smaller domain size in the case of the UFG Cu with Al_2O_3 and, thereby, the increased volume fraction η of the distorted regions. Neglecting the dislocation density inside grain interiors, which is more than two order of magnitude smaller [5], the mean dislocation density can be expressed by Eq. (4) where η denotes the volume fraction of the distorted regions and ϱ_D is the dislocation density inside the distorted regions. If the dislocation density ϱ_D is the same in both samples, the volume fraction of the distorted regions has to be higher for UFG Cu with Al_2O_3 by the same factor as the mean dislocation density. Assuming spherical grains, the volume fraction of the distorted regions is given by Eq. (5) where a is the coherent domain size and δ denotes the thickness of the distorted regions. Using the domain sizes determined by XRD (Table 2) and $\delta = 10$ nm [3], we obtained a value of η for UFG Cu with Al_2O_3 which is higher by factor of 1.4 than η for pure UFG Cu. Thus, the XRD data also indicate that the dislocation density inside the distorted regions is virtually the same in both materials.

$$\varrho = \eta \varrho_D \tag{4}$$

$$\eta = \frac{(a+\delta)^3 - (a)^3}{(a+\delta)^3} \tag{5}$$

According to our previous results on UFG Cu [5] and TEM studies of UFG Cu [4] and UFG Cu + 0.5 wt.% Al_2O_3 [12], we further assume that the dislocation density inside the distorted regions does not change with temperature and that practically unchanged distorted regions are consumed by the recrystallizing grains during the recovery process. Using these assumptions, it is possible to obtain the volume fraction of the distorted regions (η), the size of the non-distorted regions ($2R$), and the concentration of the microvoids (c_v), also in the case of saturated positron trapping. The sizes of the non-distorted regions for both specimens calculated using the model from measured PL spectra for as-prepared UFG-Cu and UFG Cu + 0.5 wt.% Al_2O_3 are given in Table 2. They correspond well to the domain size obtained by XRD, which is confirmed also by the values in Table 2, and both quantities are very closely related to the grain sizes in the specimen. It has been found that the domain size (or size of the non-distorted regions) is slightly smaller than the mean grain size determined by TEM [4, 5]. This results from the fact that the former quantities are correlated with different dislocation densities inside the grains and inside the distorted layers along GBs, whereas the grain size is determined from different contrasts for grains and GBs in the TEM image; for more detailed discussion, see Refs. [4, 5]. The difference between these quantities obviously decreases with annealing temperature due to a change of non-equilibrium GBs to the equilibrium ones. One can see from the Table 2 that UFG Cu + 0.5 wt.% Al_2O_3 exhibits smaller grain sizes than pure UFG Cu, despite of the same preparation procedure.

Evolution of structure with temperature

It is known that the mean positron lifetime $\bar{\tau}$ is a robust parameter which does not depend on the number of components to which a PL spectrum is decomposed. Therefore, we used this parameter for the description of changes occuring during isochronal annealing of UFG Cu and UFG Cu + 0.5 wt.% Al_2O_3. The dependence of $\bar{\tau}$ on the annealing temperature for both the samples is shown in Fig. 4. One can see that $\bar{\tau}$ for UFG Cu + 0.5 wt.% Al_2O_3 is remarkably higher than for pure UFG Cu due to higher number of microvoids in UFG Cu + 0.5 wt.% Al_2O_3. Moreover, in contrast to the pure UFG Cu, it continuously increases up to 400°C. In the case of the pure UFG Cu, $\bar{\tau}$ starts to decrease at 370°C, whereas in the case of UFG Cu + 0.5 wt.% Al_2O_3 the decrease starts at 400°C, *i.e.* at a similar temperature.

The temperature dependence of the lifetimes of individual components for UFG Cu and UFG Cu + 0.5 wt.% Al_2O_3 is shown in Fig. 5; the corresponding relative intensities are plotted in Fig. 6. As one can see from Figs. 5 and 6, no free positron component was observed in UFG Cu + 0.5 wt.% Al_2O_3 up to a temperature of 430°C. On the other hand, a continuous increase of intensity I_1 takes place in pure UFG Cu from 160°C (Fig. 6), and strong increase of this intensity occurs above 370°C. This difference clearly indicates that the recovery of defects is much more pronounced in pure UFG Cu compared to UFG with Al_2O_3, *i.e.* the presence of the Al_2O_3 nanoparticles prevents the recovery of defects.

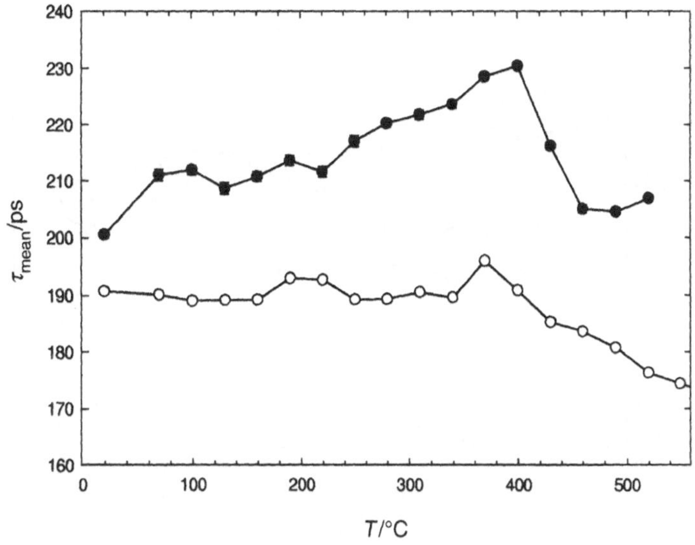

Fig. 4. Temperature dependence of the mean positron lifetime for UFG Cu (open circles) and UFG Cu + 0.5 wt.% Al_2O_3 (full circles)

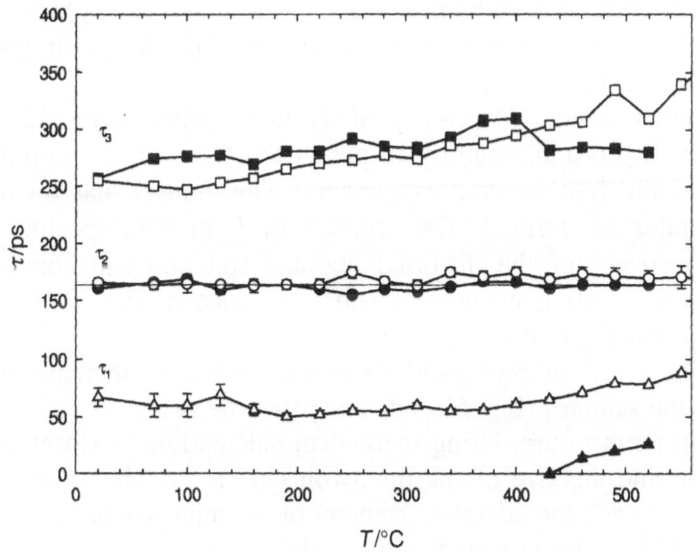

Fig. 5. Lifetimes of the components determined from PL spectra of UFG Cu (open symbols) and UFG Cu with 0.5 wt.% Al_2O_3 (full symbols) as a function of annealing temperature

The behaviour of the lifetimes τ_2 and τ_3 with temperature is practically the same for both samples (Fig. 5). Thus, the same kind of defects, *i.e.* dislocations in the distorted regions and the microvoids inside the grains, are present in both samples. The difference between the two samples consists in different concentration of these defects as well as different temperature intervals of their recovery.

The lifetime τ_2 of positrons trapped at dislocations exhibits no change with temperature. The intensity of this component starts to decrease from 160°C in pure

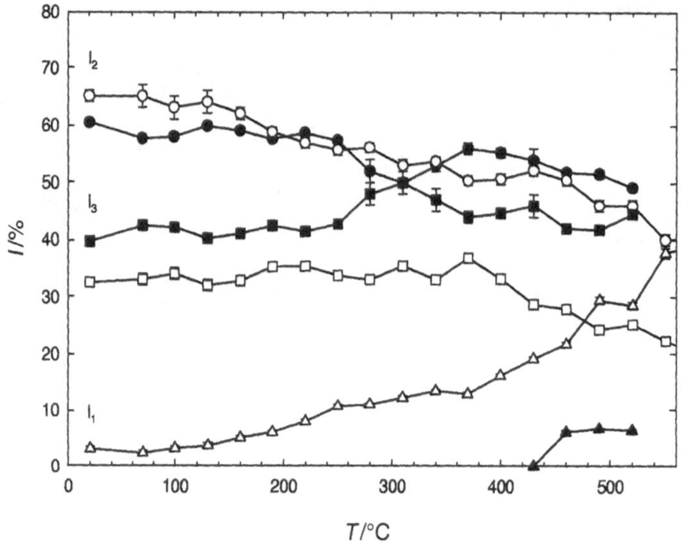

Fig. 6. Relative intensities of the components determined from PL spectra of UFG Cu (open symbols) and UFG Cu with 0.5 wt.% Al_2O_3 (full symbols) as a function of annealing temperature

UFG Cu and is correlated with an increase of the intensity I_1 of the free positron component. TEM investigations revealed an onset of the grain growth in this temperature region [4].

On the other hand, a decrease of I_2 takes place from 250°C in UFG Cu + 0.5 wt.% Al_2O_3 and, contrary to pure UFG Cu, is not accompanied by an appearance of the free positron component. This means that no recrystallized defect-free grains are formed. The decrease of I_2 in both specimens obviously reflects the recovery of the distorted regions, and one can conclude that the presence of the Al_2O_3 nanoparticles shifts the start of this recovery by about 100°C to higher temperatures.

The lifetime τ_3 of the microvoid exhibits a continuous increase with temperature for both the samples (Fig. 5), indicating that the mean size of the microvoids increases with temperature. Using theoretical calculations performed in Ref. [5], one can determine the size of the microvoids from τ_3. The temperature dependences of the size and the effective diameter of the microvoids for both specimens are plotted in Fig. 7. One can see that the initial size of the microvoids corresponds to 5 vacancies and increases to about 10 at 450°C in pure UFG Cu. The increase of size of the microvoids is less pronounced in UFG Cu + 0.5 wt.% Al_2O_3.

The relative intensity I_3 exhibits a decrease from about 370°C in pure UFG Cu. In the case of UFG Cu + 0.5 wt.% Al_2O_3 it increases from 250°C to 300°C because of the normalization of the relative intensities $I_1 + I_2 + I_3 = 100\%$ and the fact that in this temperature range $I_1 = 0\%$. The increase of I_3 is, therefore, due to the corresponding decrease of I_2 and does not indicate an increase of the number of the microvoids in this temperature interval. A determination of the concentration of the microvoids must be performed using the positron diffusion model. However, after the increase, I_3 decreases again, accompanied by the appearance of the free positron component I_1.

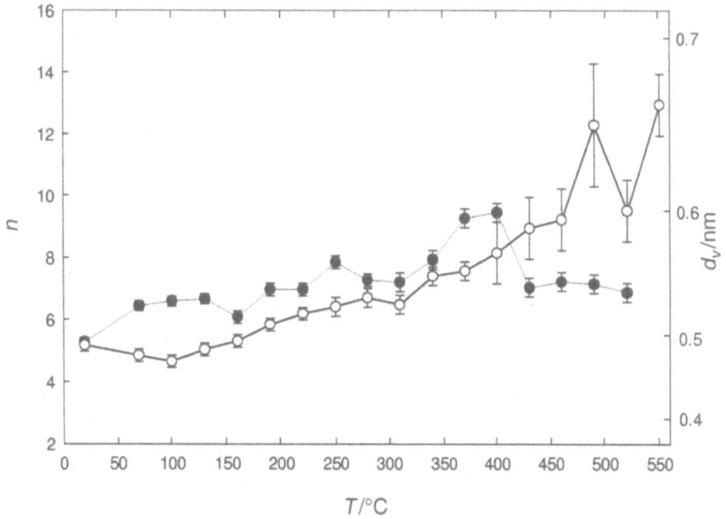

Fig. 7. Size n (in number of vacancies) of the microvoids as a function of annealing temperature for UFG Cu (open symbols) and UFG Cu + 0.5 wt.% Al_2O_3 (full symbols); the size was calculated using results from Ref. [5]; the effective diameter d_v of the microvoids is shown on the right y-axis

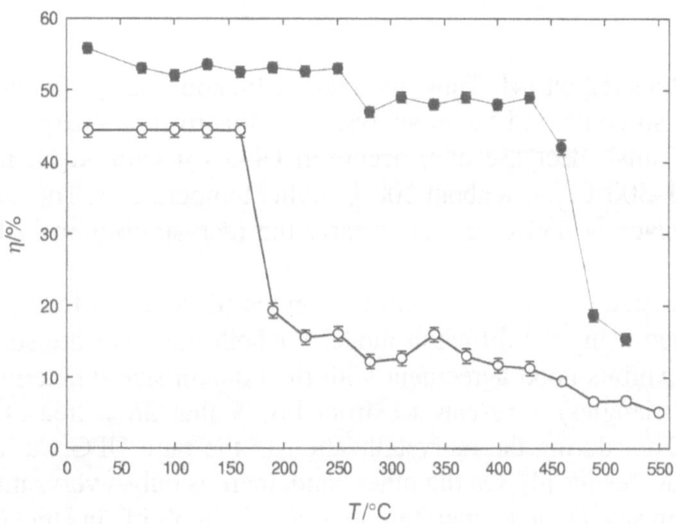

Fig. 8. Temperature dependence of the volume fraction of the distorted regions for pure UFG Cu (open circles) and UFG with 0.5 wt.% Al_2O_3 (full circles)

A further elaboration of PL spectra was performed using the diffusion model of positron behaviour in UFG materials [5, 10]. The temperature dependence of the volume fraction of the distorted regions η is plotted in Fig. 8. The higher initial volume fraction of the distorted regions in UFG Cu with Al_2O_3 is obviously due to the smaller grain size as compared to pure UFG Cu (Table 2) and the estimation performed above. In the case of pure UFG Cu, η radically decreases from 160°C to 250°C; TEM investigations have revealed that rapid grain growth takes place in

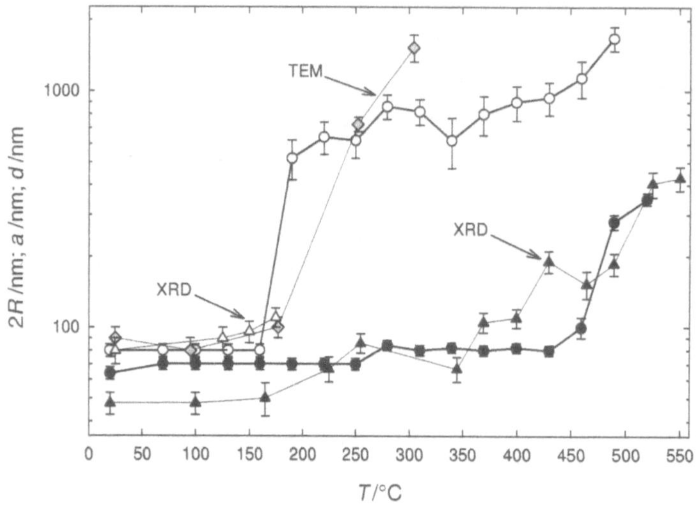

Fig. 9. Temperature dependence of the linear size of the non-distorted regions, $2R$, as determined from PL data using the diffusion model [5]; full circles: UFG Cu with Al_2O_3, open circles: pure UFG Cu; domain size a as determined by XRD for UFG Cu with Al_2O_3 (full triangles) and pure UFG Cu (open triangles); the temperature dependence of the mean grain size obtained by TEM for pure UFG Cu [4] is also included in the figure (gray diamonds)

this temperature region [4]. Thus, the recrystallization occurs in this temperature interval, as also confirmed by an analysis of the activation energy of this process [23]. Only a small decrease of η occurs in UFG Cu with Al_2O_3 in temperature region of 250–300°C, *i.e.* at about 100°C higher temperatures. The radical decrease of η taking place at 430°C reflects clearly the recrystallization in UFG Cu with Al_2O_3.

The mean sizes of the non-distorted regions (dislocation-free grain interiors), $2R$, determined using the diffusion model for both materials are shown in Fig. 9. Clearly, $2R$ exhibits good agreement with the domain size a determined by XRD (Fig. 9, full triangles). One can see from Fig. 9 that $2R$ increases rapidly from 80 nm to 500 nm during the recrystallization in the pure UFG Cu, corresponding well with TEM results [4]. On the other hand, there is only a very small increase of the mean grain size in the temperature region of 250–300°C in UFG Cu + 0.5 wt.% Al_2O_3. The most probable explanation for this fact is that only a small fraction of grains exhibits growth, *i.e.* only a few recrystallized grains appear in the deformed matrix. The recrystallization takes place from 430°C in UFG Cu with Al_2O_3 (Fig. 9) and leads to a rapid increase of grain size in the whole volume, accompanied by a decrease of the volume fraction of the distorted regions (Fig. 8). It is consistent with recent TEM results [12] which have shown that the fraction of the recrystallized grains at 400°C is not higher than 5–6%; after annealing at 600°C, it amounts to 90%. The coherent domain size lies well below 100 nm up to 400°C (Fig. 9). The mean grain size determined by TEM in Ref. [12] is still below 100 nm at 400°C in UFG Cu + 0.5 wt.% Al_2O_3, whereas it is about 1 μm, *i.e.* one order of magnitude higher, in pure UFG Cu. Thus, in agreement with Ref. [12], we can conclude

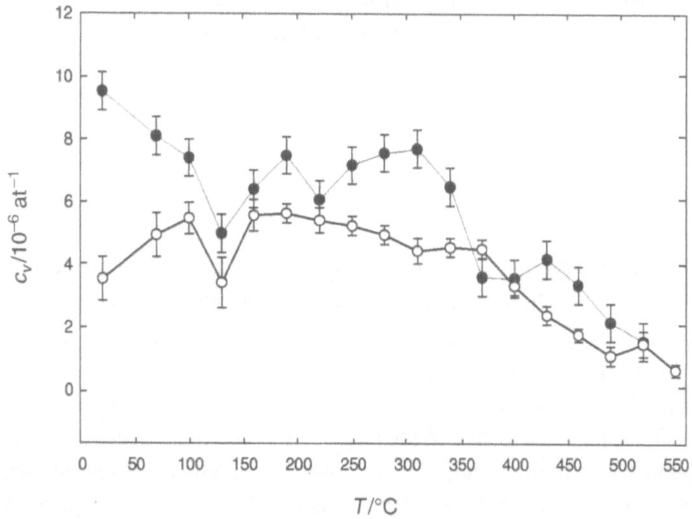

Fig. 10. Temperature dependence of concentration of the microvoids calculated from PL data using the diffusion model [5]; full symbols: UFG Cu with Al_2O_3, open symbols: pure UFG Cu

that the grain growth is substantially prevented by the presence of the Al_2O_3 nanoparticles in UFG Cu.

The concentration of the microvoids c_v can be obtained from the positron diffusion model [5, 10] and is plotted in Fig. 10. It should be noted that PL spectroscopy represents a unique tool for the investigation of the behaviour of the microvoids which are hardly detectable by other techniques. One can see from Fig. 10 that the concentration of the microvoids is higher in UFG Cu with Al_2O_3 than in pure UFG Cu. At low temperatures (below 150°C), c_v exhibits some decrease in UFG Cu + 0.5 wt.% Al_2O_3; no such behaviour occurs in pure UFG Cu. During the recrystallization in pure UFG Cu, c_v only moderately decreases. A rapid decrease starts at 370°C, *i.e.* in already recrystallized UFG Cu. In the case of UFG Cu + 0.5 wt.% Al_2O_3, c_v exhibits a decrease from 300°C, *i.e.* prior to recrystallization, and a further decrease takes place above 400°C during recrystallization. In general, the behaviour of the microvoids with annealing temperature is relatively similar in both specimens, despite of completely different temperature regions of the recrystallization.

Conclusions

The results obtained in the present work can be summarized into the following items:

(*i*) The decrease of the mean dislocation density, which is caused by the decrease of the volume fraction of the distorted regions along GBs, is shifted to significantly higher temperatures in UFG Cu with 0.5 wt.% Al_2O_3. The recrystallization occurs in the temperature region of 160–250°C in pure UFG Cu, whereas in UFG Cu with 0.5 wt.% Al_2O_3 it starts above 400°C. The mean grain size in UFG Cu with 0.5 wt.% Al_2O_3 remains below 100 nm up to 400°C.

(*ii*) UFG Cu with 0.5 wt.% Al_2O_3 contains a higher number of microvoids than pure UFG Cu. The behaviour of the microvoids with annealing temperature is similar in both materials.

Experimental

Specimens

Pure copper of purity 99.99% and a mixture of copper (99.9%) with 0.5 wt.% Al_2O_3 nanoparticles (GlidCop Al-15) were studied. In order to fabricate the UFG structure, the specimens were subjected to HPT at 6 GPa and room temperature. The true logarithmic strain can be expressed as $e = \ln(\theta r/l)$, where θ is the rotation angle in radians, and r and l are the radius and thickness of the disk, respectively [3]. In our case, $e = 7$, which corresponds to 7 rotations. The HPT technique has been described in detail in Refs. [1–3]. The microstructure of the as-prepared state of the specimens was investigated by PL and XRD spectroscopy. Subsequently, the specimens were subjected to isochronal annealing. The temperature step was choosen as 30°C, and specimens were annealed for 30 min at each temperature, *i.e.* the corresponding effective heating rate was 1°C/min. The annealing was carried in an oil base thermostat up to 250°C and in a vertical furnace with a protective argon atmosphere above this temperature. Each annealing step was finished by quenching with water of room temperature. Both PL and XRD measurements were performed at room temperature.

Positron lifetime spectroscopy

A PL spectrometer as described in Refs. [24, 25] was employed in the present work. A ^{22}Na positron source (activity: ~1.3 MBq) sealed between two mylar foils (thickness: 2 μm) was used. The timing resolution of the spectrometer was 160 ps (FWHM) for ^{22}Na at a typical coincidence counting rate of ~80 s^{-1}. At least 10^7 counts were collected in each PL spectrum. The measured spectra were decomposed by employing the maximum-likelihood procedure [26].

X-Ray diffraction

X-Ray studies were carried out using XRD7 and HZG4 (Seifert-FPM) powder diffractometers (CuK_α radiation filtered with a nickel foil, *Soller* slits placed in the diffracted beam). XRD profiles were fitted with the *Pearson* VII function by the program DIFPATAN [27]. XRD line broadening was evaluated by integral breadths (β) and FWHMs in terms of the WH plots (β *vs.* sinθ). The correction for instrumental broadening was performed with the aid of a NIST LaB_6 standard and the *Voigt* function method. Then, the modified *WH* method was used for the determination of coherent domain size and dislocation density (lattice microstrains).

Acknowledgements

This work was supported by the Czech Ministry of Education, Youth and Sport (project COST OC 523.50). The residency of *Jakub Čížek* at the University of Göttingen is supported by the *Alexander von Humboldt* Foundation.

References

[1] Valiev RZ (1997) Mat Sci Eng A **234–236**: 59
[2] Valiev RZ (1995) Nanostructured Mater **6**: 73
[3] Valiev RZ, Islamgaliev RK, Alexandrov IV (2000) Progress in Materials Science **45**: 103

[4] Islamgaliev RK, Chmelík F, Kužel R (1997) Mat Sci Eng A **237**: 43

[5] Čížek J, Procházka I, Cieslar M, Kužel R, Kuriplach J, Chmelík F, Stulíková I, Bečvář F, Melikhova O, Islamgaliev RK (2002) Phys Rev B (in press)

[6] Alexandrov IV, Zhang K, Kilmametov AR, Lu K, Lu BZ, Valiev RZ (1997) Mat Sci Eng A **234–236**: 331

[7] Valiev RZ, Musalimov RS, Tsenev NK (1989) Phys Stat Sol (A) **115**: 451

[8] Valiev RZ, Alexandrov IV, Islamgaliev RK (1998) In: Chow GM, Noskova NI (eds) Nanocrystalline Materials: Science and Technology, NATO ASI. Kluwer, p 121

[9] Valiev RZ, Islamgaliev RK (1998) In: Ghosh AK, Bieler TR (eds) Superplasticity and Superplastic Forming 1998. The Minerals, Metals and Material Society, p 117

[10] Čížek J, Procházka I, Vostrý P, Chmelík F, Islamgaliev RK (1999) Acta Phys Pol A **95**: 487

[11] Islamgaliev RK, Amirkhanov NM, Bucki JJ, Kurzydlowski KJ (2000) In: Lowe TC, Valiev RZ (eds) Investigations and Applications of Severe Plastic Deformation, NATO Science Series 3: High Technology 80. Kluwer, p 297

[12] Buchgraber W, Islamgaliev RK, Kolobov YUR, Amirkhanov NM (1999) In: NATO ARW. Moscow August 2–9, 1999

[13] Kužel R, Čížek J, Procházka I, Chmelík F, Islamgaliev RK, Amirkhanov NM (2001) In: METAL 2001. Ostrava, Tanger Ltd (CD ROM)

[14] Hautojärvi P, Corbel C (1995) In: Dupasquier A, Mills AP (eds) Proceedings of the International School of Physics "Enrico Fermi", Course CXXV. IOS Press, Varenna, p 491

[15] McKee BTA, Saimoto S, Stewart AT, Scott MJ (1974) Can J Phys **52**: 759

[16] de Lima AP, Lopes Gil C, Martins DR, Ayres de Campos N, Menezes LF, Fernandes JV (1987) In: Dlubek G, Brümmer O, Brauer G, Hennig K (eds) Proceedings od European Meeting on Positron Studies of Defects, vol. 2, part 1. Martin-Luther-Universität Halle-Wittenberg, Wernigerode, p C1

[17] Dupasquier A, Romero R, Somoza A (1993) Phys Rev B **48**: 9235

[18] PDF-2, Powder Diffraction Pattern Database, ICDD (International Centre for Diffraction Data), record number 04-0836

[19] Ungár T, Dragomir I, Revesz I, Borbely A (1999) J Appl Cryst **32**: 992

[20] Ungár T, Tichy G (1999) Phys Stat Sol (B) **171**: 425

[21] Ungár T, Gubiza J, Ribárik G, Borbély A (2001) J Appl Cryst **34**: 298

[22] Kužel R, Čížek J, Procházka I, Chmelík F, Islamgaliev RK, Amirkhanov NM (2001) Mat Sci Forum **378–381**: 463

[23] Čížek J (2001) PhD Thesis, Charles University, Prague, Czech Republic

[24] Bečvář F, Čížek J, Lešták L, Novotný I, Procházka I, Šebesta F (2000) Nucl Instr Meth A **443**: 557

[25] Bečvář F, Čížek J, Procházka I (1999) Acta Physica Polonica A **95**: 448

[26] Procházka I, Novotný I, Bečvář F (1997) Mat Sci Forum **255–257**: 772

[27] Kužel R (1995) DIFPATAN – Program for Powder Pattern Analysis, http://www.xray.cz/priv/kuzel/difpatan

Received October 5, 2001. Accepted (revised) December 20, 2001

Vibrational Spectroscopy and Analytical Electron Microscopy Studies of Fe–V–O and In–V–O Thin Films

Angela Šurca Vuk[1], **Boris Orel**[1,*], **Goran Dražič**[2], and **Philippe Colomban**[3]

[1] National Institute of Chemistry, SL-1000 Ljubljana, Slovenia
[2] Josef Stefan Institute, SL-1000 Ljubljana, Slovenia
[3] Laboratoire de Dynamique, Interactions et Reactivite, Centre National de la Recherche Scientifique – Universite Pierre et Marie Curie, F-94320 Thiais, France

Summary. Orthovanadate ($M^{3+}VO_4$; M = Fe, In) and vanadate ($Fe_2V_4O_{13}$) thin films were prepared using sol-gel synthesis and dip coating deposition. Using analytical electron microscopy (AEM), the chemical composition and the degree of crystallization of the phases present in the thin Fe–V–O films were investigated. TEM samples were prepared in both orientations: parallel (plan view) and perpendicular (cross section) to the substrate. In the first stages of crystallization, when the particle sizes were in the nanometer range, the classical identification of phases using electron diffraction was not possible. Instead of measuring *d* values, experimentally selected area electron diffraction (SAED) patterns were compared to calculated (simulated) patterns in order to determine the phase composition. The problems of evaluating the ratio of amorphous and crystalline phases in thin films are reported.

Results of TEM and XRD as well as IR and *Raman* spectroscopy showed that the films made at lower temperatures (300°C) consisted of nanograins embedded in the dominating amorphous phase. Characteristic vibrational spectra allowed to distinguish between the different crystalline phases, since the IR and *Raman* bands showed broadening due to the decreasing particle size of the films thermally treated at lower temperatures. Vibrational analysis also showed that the electrochemical cycling of crystalline films led to spectra that were in close agreement with the spectra of the nanocrystalline films prepared at lower temperatures. The formation of a nanocrystalline structure is therefore a prerequisite for obtaining a higher charging/discharging stability of Fe–V–O and In–V–O films.

Keywords. Vanadates; Thin films; Sol-gel; Electrochromism; IR spectroscopy; Electron microscopy.

Introduction

The electrochromic properties of transition metal orthovanadates ($M^{3+}VO_4$, M = Ce, Fe, In, ...) have been extensively studied because of their potential for use as cathode materials in lithium rocking chair batteries [1–3]. Relatively little work

* Corresponding author. E-mail: boris.orel@ki.si

has been carried out on thin films of these materials, despite the fact that they exhibit electrochromism [4–6]. One of the most attractive properties of the electrochromic orthovanadates is their low optical response (below 20% T) during the insertion of Li^+ ions and electrons together with a high charge insertion/extraction capacity which makes it possible to use these films in electrochromic devices in combination with electrochromically active WO_3 or Nb_2O_5 [7].

Electrochemical studies revealed that amorphous V_2O_5 powders exhibit long-term cycling stability (> 1000 cycles), better than that observed for the corresponding crystalline V_2O_5 powders [8]. In our recent studies of In–V–O and Fe–V–O films we have shown that sol-gel synthesis permitted the production of films which exhibited, after different thermal treatments, amorphous, nanocrystalline, or well-defined crystalline structures [9–14]. One of the aims of this study is to show, by means of vibrational spectroscopy, correlations between the structure of In–V–O and Fe–V–O films prepared at low temperatures (300°C) and the structure that the initially crystalline $InVO_4$, $FeVO_4$, and $Fe_2V_4O_{13}$ films obtain after electrochemical cycling.

In addition to vibrational spectroscopy, analytical electron microscopy was also used. Due to its lateral and spatial resolution it is one of the most effective methods for studying thin films of various thicknesses and compositions. Using energy-dispersive X-ray spectroscopy (EDXS), electron diffraction, and dark-field imaging, crystalline phases with grain sizes down to a few nanometers can be characterized. Problems arise, however, when the crystal size of the phases is even smaller, *i.e.* in the region where the nanocrystalline phase is formed from the amorphous phase. The usually accepted definition [15] of an amorphous material is one for which the locations of the neighbouring atoms are defined by a probability function with probabilities lower than unity. In an amorphous material the electron scattering from areas as large as 1.5 nm can be coherent [16, 17]. The consequence of coherent scattering is that the selected area electron diffraction (SAED) patterns consist of diffuse rings (haloes), whereas in dark-field images spots with bright contrast are present. It is characteristic of the technique that the size of the bright spots increases with increasing defocus. In the case of a nanocrystalline material, the rings in SAED patterns are sharper and more complex; in dark-field images, the bright crystallites are well resolved. The identification of phases from electron diffraction patterns of nanocrystalline samples is often problematic due to the diffuse rings. In the present work we report on the identification of nanocrystalline phases in thin Fe–V–O films using a comparison of experimental and the simulated SAED patterns.

Results and Discussion

Analytical electron microscopy of films

Simulated electron diffraction 'powder' patterns were calculated using the EMS program package (Electron Microscopy Simulation program from Dr. *P. Stadelmann*, EPFL-CIME, Lausanne, Switzerland) [18]. As a result, a listing of the positions of the powder lines and the intensity of the lines was created. Intensities were calculated from the structure factors and corrected by a shape factor depending on the crystal

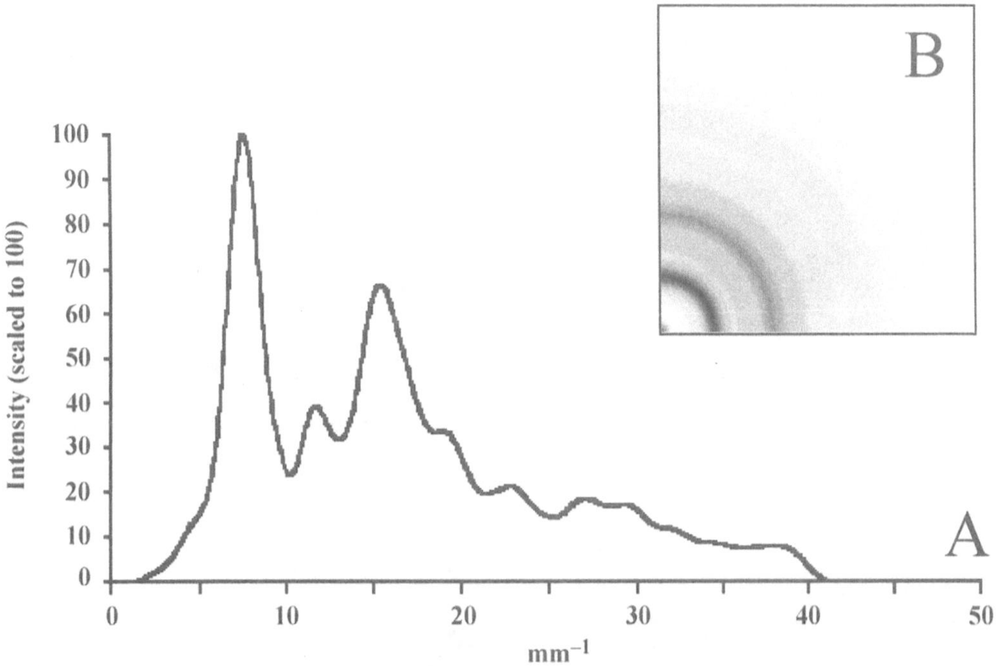

Fig. 1. A) Calculated distribution of intensities; B) their graphical representation for a triclinic $FeVO_4$ with 1 nm crystal size at 200 keV beam energy and 980 mm camera length

size. Dynamic effects were not considered in the calculations. To be able to compare the experimental and the simulated electron diffraction patterns, images were constructed from the intensity distributions. To determine the compounds present in the thin films we systematically calculated electron diffraction patterns for all possible phases for a mean crystallite size between 0.5 and 5 nm. For the simulations we used published crystal structure data for $Fe_2V_4O_{13}$ [19], $FeVO_4$ [20], $FeVO_4$-II (high-pressure form) [21], $InVO_4$-I [22], $InVO_4$-III [23], Fe_2O_3 [24], and V_2O_5 [25].

An example of a calculated intensity distribution for a triclinic $FeVO_4$ phase [20] with a crystal size of 1 nm and a corresponding simulated electron diffraction pattern are shown in Fig. 1. In this compound, three independent iron atoms are joined in a doubly bent chain of six edge-sharing iron polyhedra [20]. Two iron atoms are in a distorted octahedral and one in a distorted trigonal bipyramidal environment. The chains of iron polyhedra are joined by VO_4 tetrahedra which share corners with up to four iron polyhedra within a single chain.

A series of simulated patterns for a monoclinic $Fe_2V_4O_{13}$ phase [19] where the crystal size varied from 0.5 to 5 nm is depicted in Fig. 2. From very diffuse haloes in the pattern for 0.5 nm, which are characteristic for amorphous materials, the rings become sharper and more complicated with increasing crystal size. The Fe_2 V_4O_{13} crystal structure [19] consists of Fe^{3+} octahedra and V^{5+} tetrahedra. Iron-distorted octahedra form dimeric Fe_2O_{10} units, whereas four VO_4 tetrahedra are linked through corners in a U-shaped $(V_4O_{13})^{6-}$ polyanion. Each Fe_2O_{10} unit shares its oxygens with seven different V_4O_{13} groups. Such an arrangement creates hexagonal and narrow tetragonal empty tunnels.

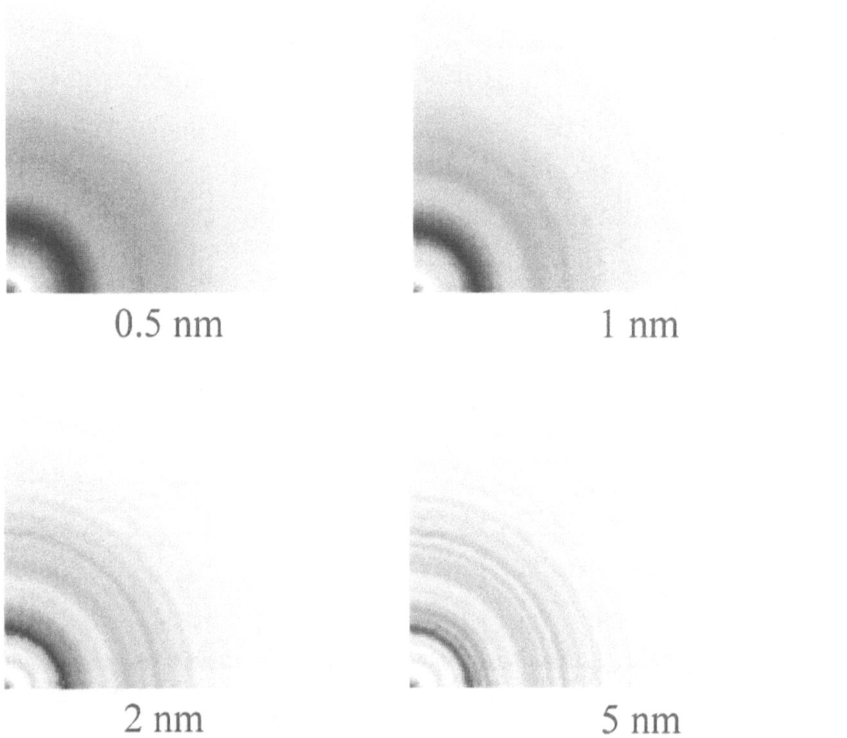

Fig. 2. Simulated SAED patterns (one quarter shown) of a monoclinic $Fe_2V_4O_{13}$ phase using crystal sizes of 0.5, 1, 2, and 5 nm

A central dark-field (CDF) TEM micrograph of a sample with the composition Fe:V = 1:2 (300°C) is displayed in Fig. 3 together with a comparison of the experimental and simulated SAED pattern. The white spots in the micrograph have a dimension of about 1 nm (the picture was collected very close to *Gauss*'s focus), and the SAED pattern was found to be quite diffuse; nevertheless, several circles of various intensity could still be resolved. The main question was: is the sample composed of just an amorphous phase or are there some crystalline phases present? The experimental diffraction pattern was compared to simulated patterns, and the best match was found in the case of the high-pressure orthorhombic form of iron vanadate ($FeVO_4$-II [21]) with a crystallite size of 1 nm (Fig. 3B). The structural features of $FeVO_4$-II [21] are one-dimensional chains of edge-sharing FeO_6 octahedra running along the *c* axis. The chains are linked through VO_4 tetrahedra.

In a thin film sample with the composition Fe:V = 1:1, which was thermally treated at 300°C, we found very similar SAED patterns but with sharper circles (Fig. 4B). In a dark-field TEM micrograph, bright areas with sizes of 5 nm and more were found (Fig. 4A). In this case the presence of nanocrystals was evident. The best match between the experimental and the simulated SAED patterns was found for the orthorhombic $FeVO_4$-II phase [21] with 5 nm particle size (Fig. 4B). Based on the similarities of the diffraction patterns and good agreement between the experimental patterns and the simulated ones we concluded that the crystalline phase was also present in the sample shown in Fig. 3 (nanometer-sized particles of

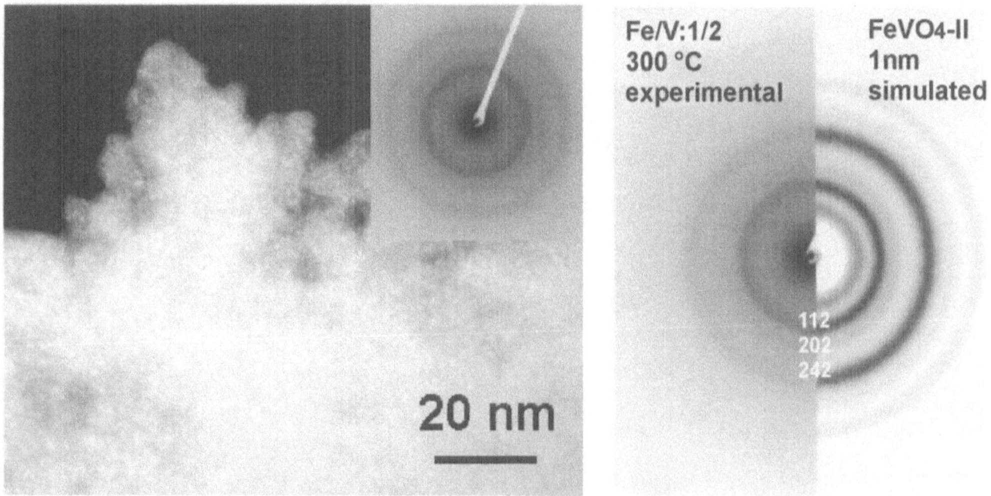

A B

Fig. 3. A) Dark-field TEM micrograph (plan view) of a thin film with a molar ratio Fe:V = 1:2 thermally treated at 300°C; B) comparison of experimental and calculated SAED patterns for an orthorhombic $FeVO_4$-II phase

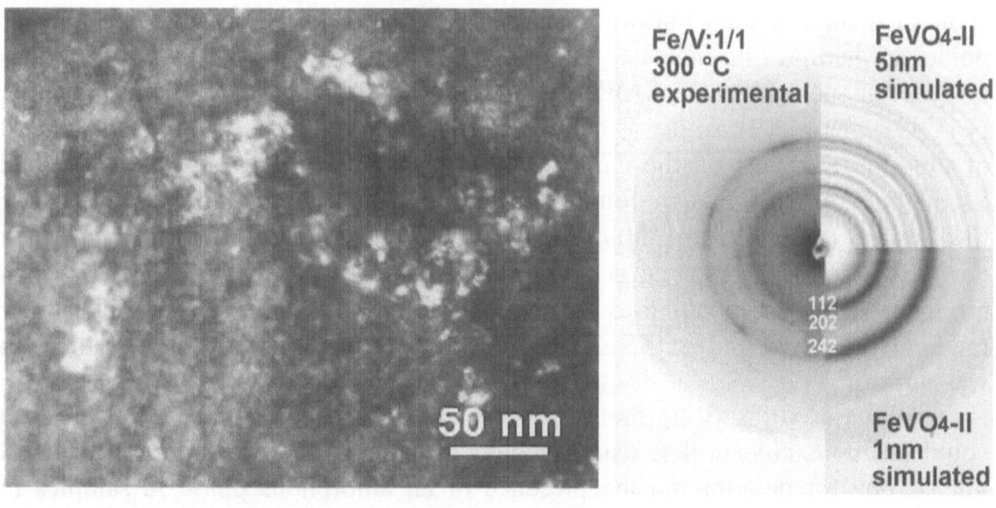

A B

Fig. 4. A) Dark-field TEM micrograph (plan view) of a thin film with a molar ratio Fe:V = 1:1 thermally treated at 300°C; B) comparison of experimental and calculated SAED patterns for an orthorhombic $FeVO_4$-II phase

orthorhombic $FeVO_4$-II). This conclusion was in agreement with the results obtained with IR spectroscopy (see below).

In Fig. 5, the central dark-field (CDF) TEM micrograph and SAED pattern of a cross-section of a Fe:V = 1:2 (400°C) thin film are shown. The estimated film thickness was close to 150 nm. Two types of crystallites were found, one with a size around 50 nm and the other with a size below 5 nm. A bimodal distribution of

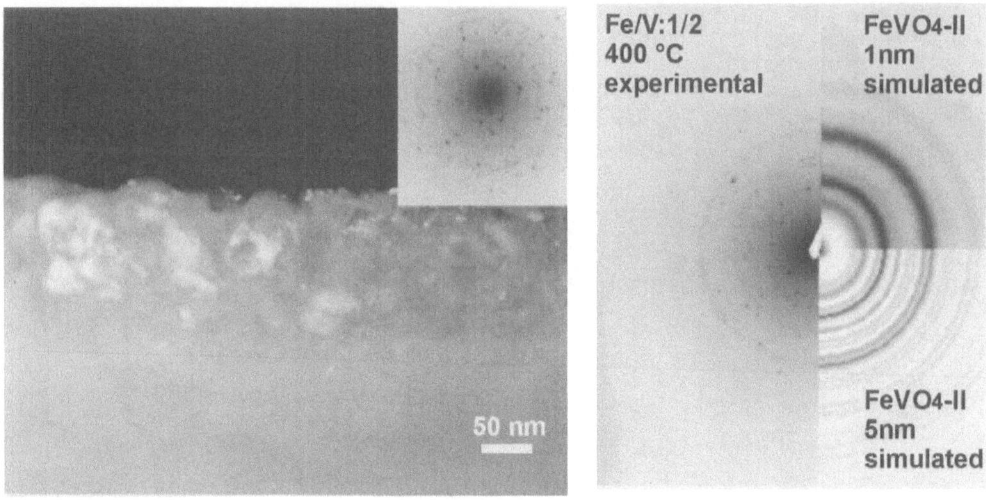

A B

Fig. 5. A) Dark-field TEM micrograph (cross section) of a thin film with a molar ratio Fe:V = 1:2 thermally treated at 400°C; B) comparison of experimental and calculated SAED patterns for a $FeVO_4$-II phase

the crystallite sizes could be seen in the SAED patterns, which besides distinctive spots also showed faint uniform rings. An EDXS analysis performed on various points on the thin film indicated the presence of Fe and V in an approximate ratio of 1:2. All spots from the SAED patterns could be indicated for the monoclinic $Fe_2V_4O_{13}$ phase [19]; so it was concluded that the larger grains in Fig. 5A consist of this phase. Based on the comparison of the experimental and the simulated SAED patterns (Fig. 5B) we found that the faint rings suggest the presence of the fine-grained (around 5 nm) orthorhombic $FeVO_4$-II phase. A TEM micrograph of the Fe:V = 1:1 film prepared at 500°C revealed grains with dimensions between 50 and 80 nm corresponding to the monoclinic $FeVO_4$ phase. These grains were randomly oriented, and EDXS showed that the chemical composition in the film was uniform.

In all cases (Figs. 3–5), the presence of an amorphous phase in the thin film could not be excluded. The usual method in conventional transmission electron microscopy for determining the presence of an amorphous phase in samples in which the major constituents are crystalline is a dark-field imaging using diffusely scattered electrons [15]. In the case of fine-grained polycrystalline samples this method is not effective because of the presence of circles in diffraction patterns. Insome cases, high-resolution electron microscopy (HRTEM) can provide the answer; however, the limitations are sample thickness, sample preparation, and microscope resolution. During the preparation of TEM samples the thinnest areas, where the HRTEM approach is effective, are most likely to have a damaged structure.

For comparison, the XRD spectra of the vanadate films prepared at the highest temperatures ($FeVO_4$ and $InVO_4$ at 500°C, $Fe_2V_4O_{13}$ at 400°C) are shown in Fig. 6. The XRD spectra revealed the most intense bands of the corresponding phases: for the films with the molar ratio Fe:V = 1:2 the monoclinic $Fe_2V_4O_{13}$ phase (JCPDS 39-08930) and for films with Fe:V = 1:1 the triclinic $FeVO_4$ phase (JCPDS

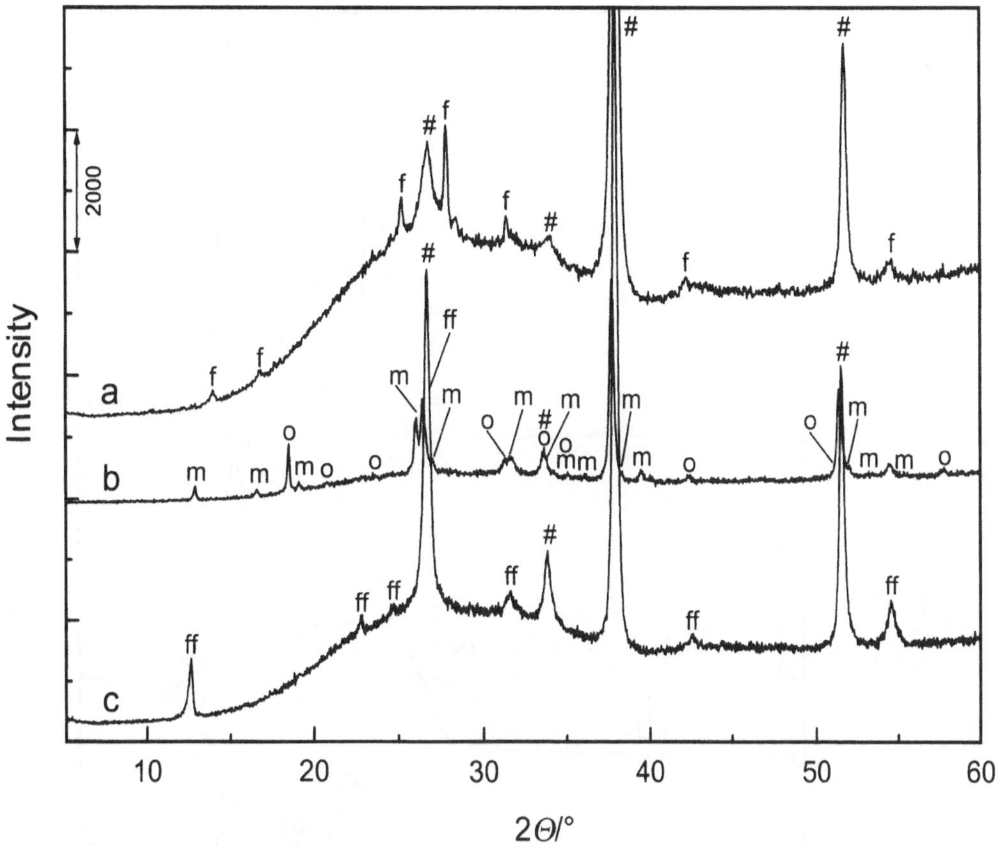

Fig. 6. XRD spectra of thin vanadate films: a) $FeVO_4$ (500°C), b) $InVO_4$ (500°C), c) $Fe_2V_4O_{13}$ (400°C); *f* denotes triclinic $FeVO_4$ phase, *ff* monoclinic $Fe_2V_4O_{13}$ phase, *m* monoclinic $InVO_4$-I phase, *o* orthorhombic $InVO_4$-III phase, and # diffraction peaks of the SnO_2/F glass substrates

38-1372). The XRD spectrum of the In:V = 1:1 film, in contrast, showed the presence of two crystalline phases: the prevailing monoclinic $InVO_4$-I (JCPDF 38-1135) phase with a grain size below 40 nm and the orthorhombic $InVO_4$-III (JCPDF 48-0898) phase. The structure of the monoclinic $InVO_4$-I phase [22] consists of compact In_4O_{16} groups of four edge-shared InO_6 octahedra. These In_4O_{16} groups are linked to each other by VO_4 tetrahedra. The structure of the orthorhombic $InVO_4$-III phase [23] is composed of chains of InO_6 octahedra which are linked together by VO_4 tetrahedra.

IR spectra of films

The IR spectra of the investigated $FeVO_4$, $InVO_4$, and $Fe_2V_4O_{13}$ films are presented in Figs. 7–9. The spectra of the films prepared at lower temperatures are characterized by broad bands which confirm the findings of the electron microscopic investigations (Figs. 1–5) – a dominating amorphous phase in which the nanograins are embedded. With increasing temperature the number of IR bands also increases, as does their sharpness. Thermal treatment at 500°C ($FeVO_4$, $InVO_4$) and 400°C ($Fe_2V_4O_{13}$) leads to the formation of crystalline films. For all

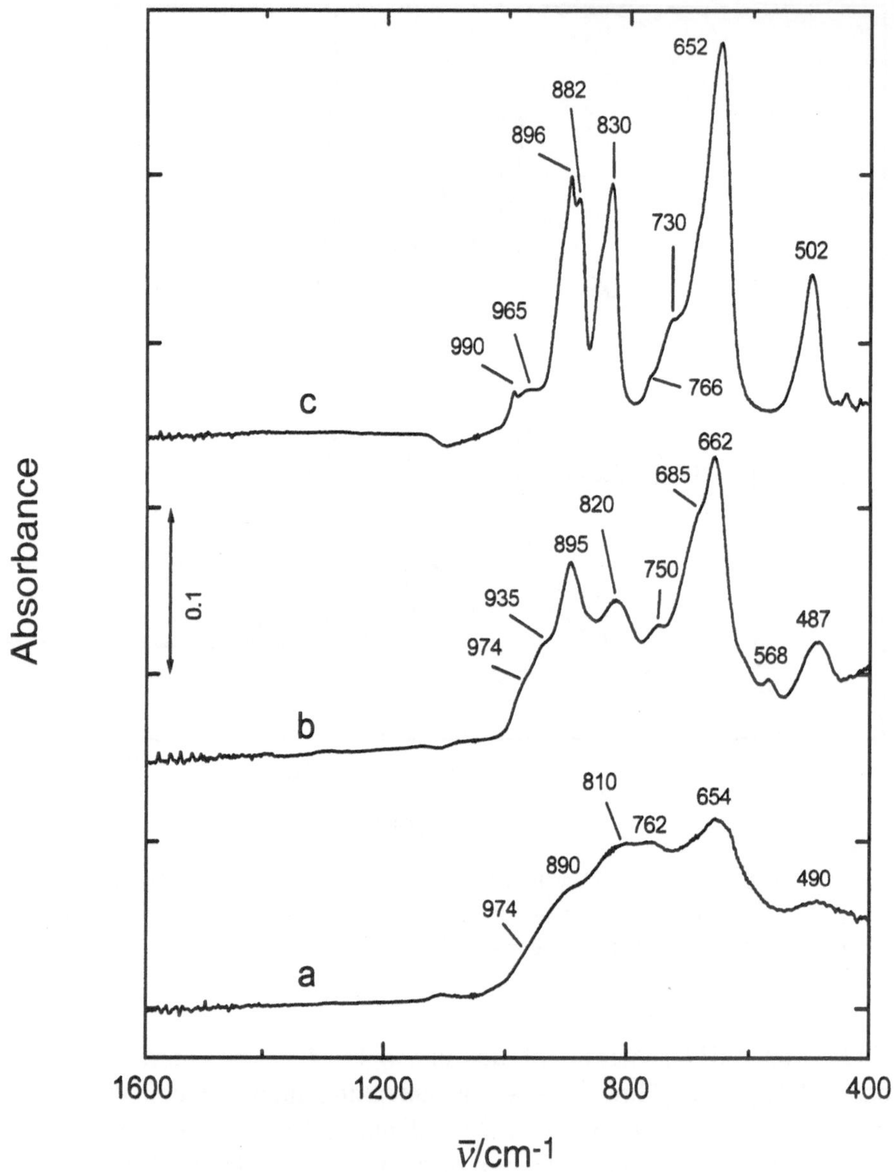

Fig. 7. IR absorbance spectra of $FeVO_4$ films thermally treated for 1 h at a) 300°C, b) 400°C, c) 500°C

investigated compounds the IR bands can be assigned to three or four major regions (Table 1) depending on their structure: V–O terminal stretching (1050 to $\sim 850\,cm^{-1}$), bridging V–O$\cdots$*M* (*M*=In, Fe), V$\cdots$O$\cdots$*M* and V–O–V stretching (~ 850 to $\sim 550\,cm^{-1}$), V–O–V deformation and Fe–O stretching ($< 550\,cm^{-1}$) [9–14, 26–29]. This assignment was made on the basis of the assignment of lead vanadate glasses according to *Hayakawa et al.* [26], who assumed that the glasses consist of $(VO_3)_n$ chains of corner-sharing VO_4 groups with V–O bonds of different strengths. In orthovanadates, VO_4 groups do not form chains among themselves and bridging V–O$\cdots$Fe and V$\cdots$O$\cdots$Fe modes with stronger or weaker bonds appear in the region between 500 and $850\,cm^{-1}$. In contrast to $InVO_4$ and $FeVO_4$ orthovanadates, the bridging V–O–V stretching is present

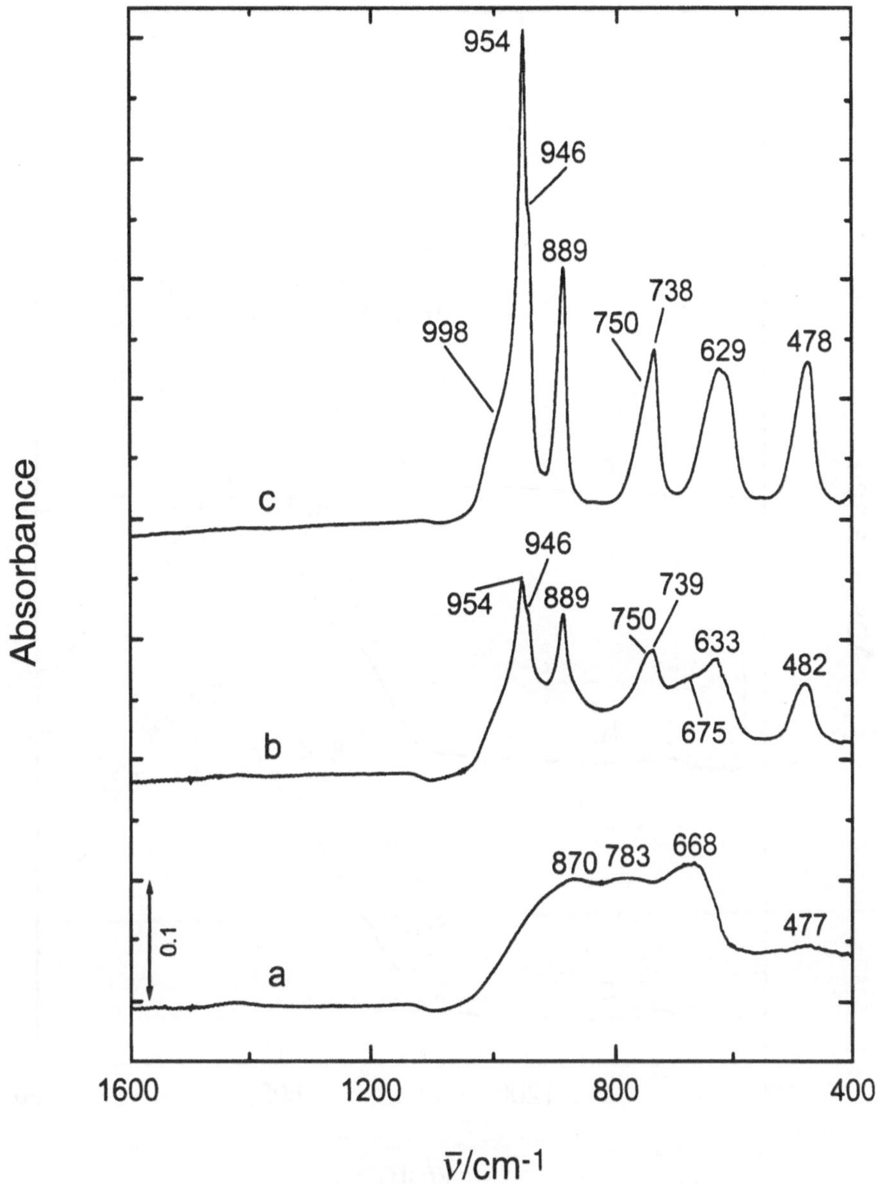

Fig. 8. IR absorbance spectra of $InVO_4$ films thermally treated for 1 h at a) 300°C, b) 400°C, c) 500°C

in the IR spectra of $Fe_2V_4O_{13}$ films due to the $(V_4O_{13})^{6-}$ anions in the crystalline structure [19].

VO_4^{3-} ions in aqueous solutions exhibit two IR active modes [30]: the totally symmetric (υ_1, A symmetry) mode at 824 cm^{-1} and the asymmetric (υ_3, F_3 symmetry) mode at 790 cm^{-1}. In condensed structures, due to the decrease in the site symmetry, the shift of the IR bands to higher frequencies and the splitting of the triply degenerated υ_3 mode are expected. The symmetric υ_1 stretching appears, for example, at 915 cm^{-1}, and the split υ_3 modes at 890, 770, and 680 cm^{-1} for $InVO_4$ and $TlVO_4$ [28]. The presence of more than three separate bands in the υ_3 stretching region indicates strong coupling between different VO_4 tetrahedra and M^{3+}–O (M = Fe, Al, Cr, ...) polyhedra in the unit cell. It is known, however, that

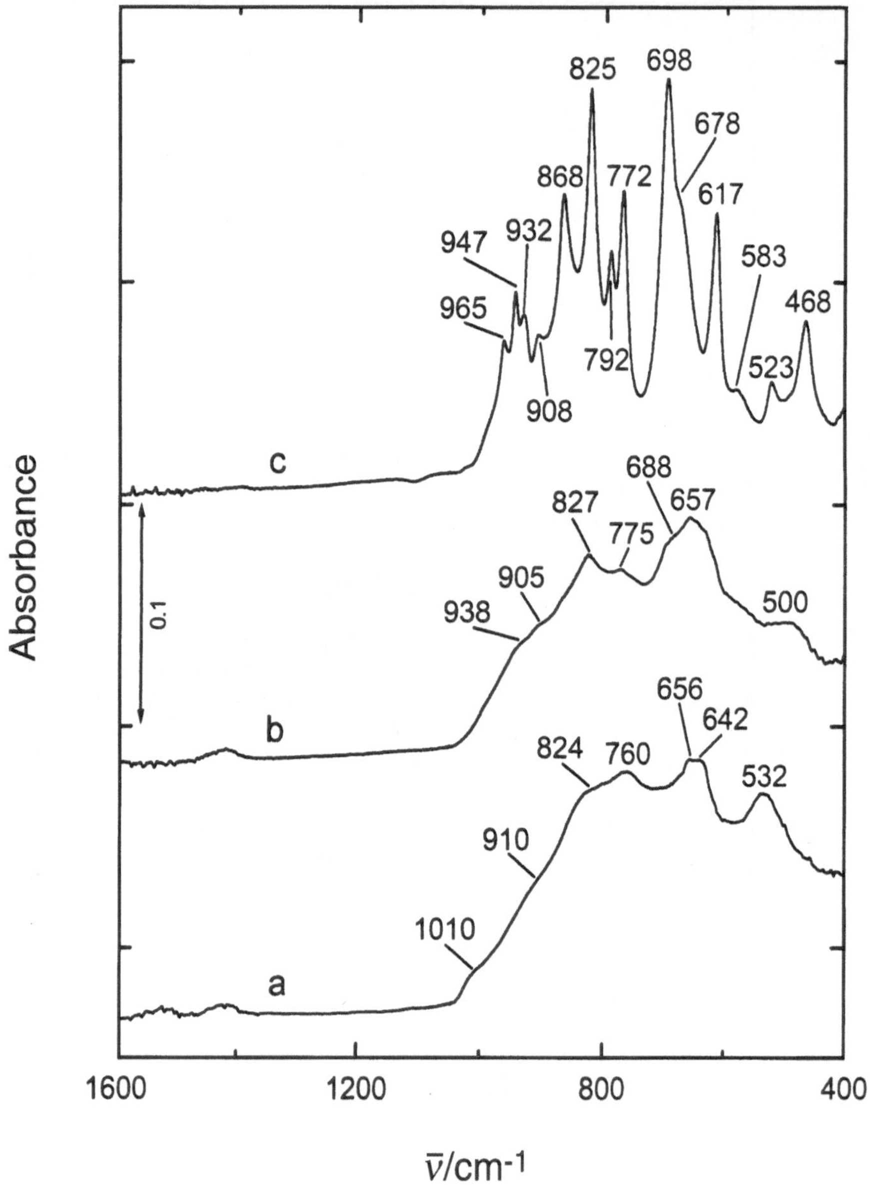

Fig. 9. IR absorbance spectra of $Fe_2V_4O_{13}$ films thermally treated for 1 h at a) 200°C, b) 300°C, c) 400°C

from pure V_2O_5 [25] to ortovanadates [20–23] a trend of longer chemical bonds and a lower coordination number of the vanadium with respect to the oxygen is observed [27]. In the IR spectra of the orthovanadates [28, 29] the intense bands around 1000 cm^{-1}, signifying the short isolated V=O bonds, are not formed. The symmetry of VO_4^{3-} groups in orthovanadates decreases due to condensation effects, resulting in an increase in the number of absorption bands in the IR spectra.

We can conclude that the crystallization of Fe–V–O and In–V–O films proceeds through different stages. In the first stage (temperature <200°C), the structure consists of an amorphous phase with vanadium polyhedra and metal ions (M = Fe, In) probably bonded on the interstitial sites (isolated V=O bonds) [27]. During the second stage (~300°C), nanograins with $FeVO_4$-II or $InVO_4$-III

Table 1. Assignment of IR bands (cm^{-1}) of $InVO_4$, $FeVO_4$, and $Fe_2V_4O_{13}$ films

Assignment	Range	300°C	400°C	500°C
$InVO_4$				
V–O terminal stretching	1000–850			998
			954	954
			946	946 sho (III)
		870	889	889
Bridging V–O···In stretching	850–700	783		
			750	750
			739	738
Brinding V···O···In stretching	700–550	668	675	
			633	629
V–O–V deformation	< 550	477	482	478
$FeVO_4$				
V–O terminal stretching	1050–880	974	974	990
		890	935	965
			895	896
				882
Bridging V–O···Fe stretching	880–700	810	820	830
		762	750	766
			685	730
Mixed bridging V–O···Fe and V···O···Fe stretchings	700–550	654	662	652
			568	
V–O–V deformation, Fe–O stretching	< 550	490	487	502
$Fe_2V_4O_{13}$				
V–O terminal stretching	1050–800	1010		965
				947
			938	932
		910	905	908
				868
		824	827	825
Mixed bridging V–O–V and V–O···Fe stretching	800–600			792
		760	775	772
			688	698
		656	657	678
		642		
				617
V–O–V deformation, Fe–O stretching	< 600			583
		532	500	523
				468

structure and dimensions of 1 nm (Fe:V = 1:2, In:V = 1:1) or 5 nm (Fe:V = 1:1) are formed. The outer parts of these nanograins are probably terminated with vanadium polyhedra. At higher temperatures (400 and 500°C) the films become crystalline, but the dimensions of the grains are different (from below 40 nm to 80 nm).

Ex situ IR absorbance spectroelectrochemical measurements

With this study we wanted to generalize the IR spectroscopic behaviour of orthovanadate ($M^{3+}VO_4$, M = Fe, In) and vanadate ($Fe_2V_4O_{13}$) thin films that were thermally treated at high temperatures (500°C for orthovanadates, 400°C for $Fe_2V_4O_{13}$) during electrochemical cycling. Prior to *ex situ* IR absorbance measurements, the films were charged or discharged in 1 *M* $LiClO_4$ in propylene carbonate (*PC*) using chronopotentiometry (i = constant) or chronocoulometry (E = constant). Details of the electrochemical charging/discharging are described in Table 2. In Figs. 10–12 (curves a, b) we first present the spectra of $FeVO_4$, $InVO_4$, and $Fe_2V_4O_{13}$ films charged/discharged to −0.04 or $-0.06\,mC \cdot cm^{-2} \cdot nm^{-1}$, the charging range in which the intercalation/deintercalation reaction of Li^+ ions is reversible. The second pair of spectra in Figs. 10–12 (curves c, d) shows the IR spectra of highly charged/discharged films which indicate large structural changes and amorphization.

The basic question when interpreting the spectral changes of charged/discharged films is the assessment of interactions between the intercalated Li^+

Table 2. Intercalation properties of crystalline $InVO_4$, $FeVO_4$, and $Fe_2V_4O_{13}$ films during *ex situ* IR absorbance measurements

Film	T_h/°C	d/nm	Charging technique	E/V	$i/\mu A \cdot cm^{-2}$	t_c/s
$InVO_4$	500	230	CE	–	22.3	448
						1792
$FeVO_4$	500	90	CE	–	35.0	143
					32.1	936
$Fe_2V_4O_{13}$	400	70	CC	−1.50	–	120
				−3.00	–	480

Film	$Q_{ins}/mC \cdot cm^{-2}$	$Q_{ins} \cdot d^{-1}/mC \cdot cm^{-2} \cdot nm^{-1}$	$\rho/g \cdot cm^{-3}$	x	x'
$InVO_4$	−10.0	−0.04	4.61	0.23	–
	−40.0	−0.17		0.90	
$FeVO_4$	−5.0	−0.06	3.65	0.27	–
	−30.0	−0.33		1.62	
$Fe_2V_4O_{13}$	−4.3	−0.06	3.12	0.27	1.07
	−23.3	−0.33		1.44	5.77

T_h: temperature of thermal treatment; d: thickness of films; E: charging potential; i: current density; t_c: time of charging; Q_{ins}: charge density; $Q_{ins} \cdot d^{-1}$: charge density per film thickness; ρ: density of crystalline phases; x: intercalation coefficient per V atom; x': molar ratio of Li in $Li_xFe_2V_4O_{13}$; all films were thermally treated for 1 h; charging/discharging was performed using chronopotentiometry (CE) or chronocoulometry (CC)

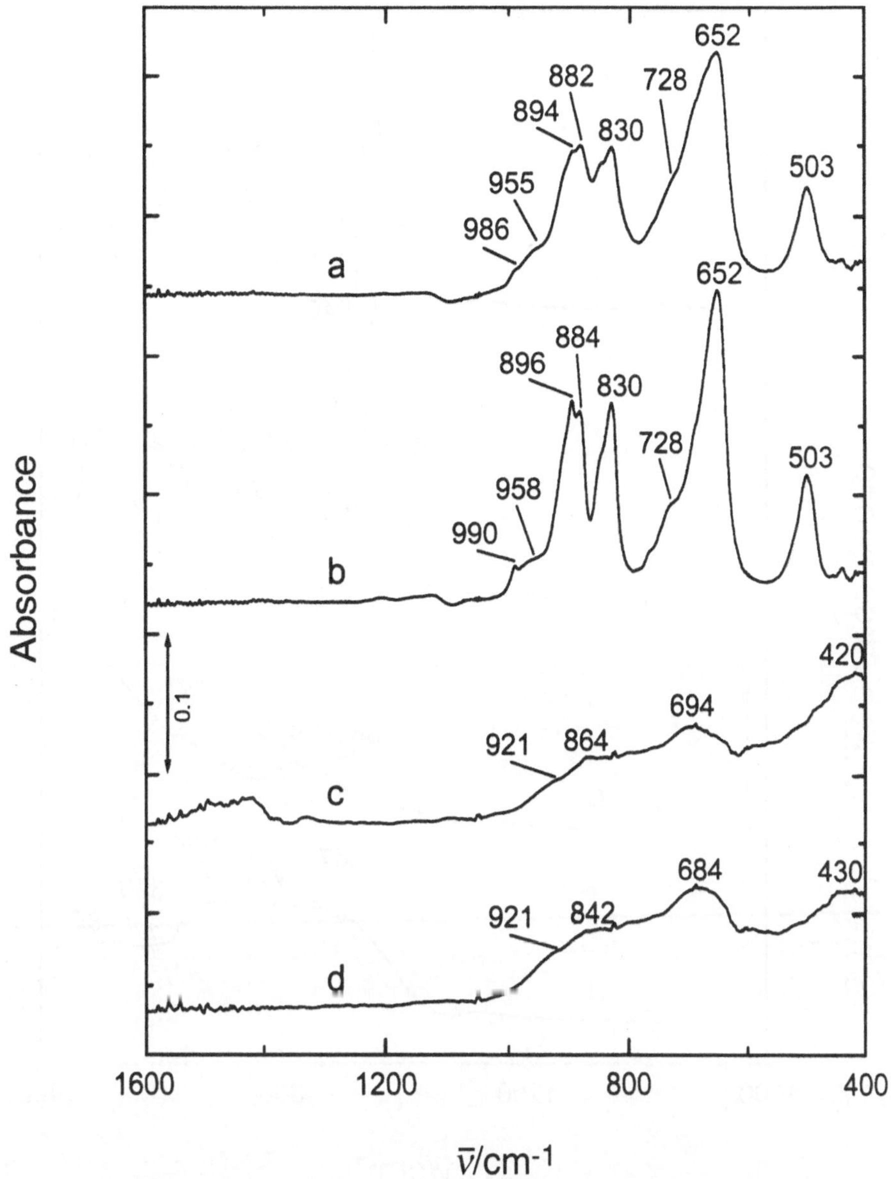

Fig. 10. *Ex situ* IR absorbance spectra of a $FeVO_4$ film (500°C, 1 h): a) charged to $-0.06\,mC \cdot cm^{-2} \cdot nm^{-1}$, b) discharged from $-0.06\,mC \cdot cm^{-2} \cdot nm^{-1}$, c) charged to $-0.33\,mC \cdot cm^{-2} \cdot nm^{-1}$, d) discharged from $-0.33\,mC \cdot cm^{-2} \cdot nm^{-1}$

ions and the film network. The strength of interactions is responsible either for only small changes in the film structure or for the amorphization of the structure. In the former case, the structural change is topotactic, and the atoms return to their original positions after discharging. For amorphous films, IR spectroscopy is a powerful technique for obtaining information about the groups of atoms that change their redox state or about the atoms that interact with the inserted Li^+ ions. Without having the results of the normal coordinate treatment at hand (rarely available for vanadates [28]), the electrochemically induced changes in films can be explained at least quantitatively by comparing the IR spectra of films in initial, intercalated, and deintercalated states to the spectra of model compounds.

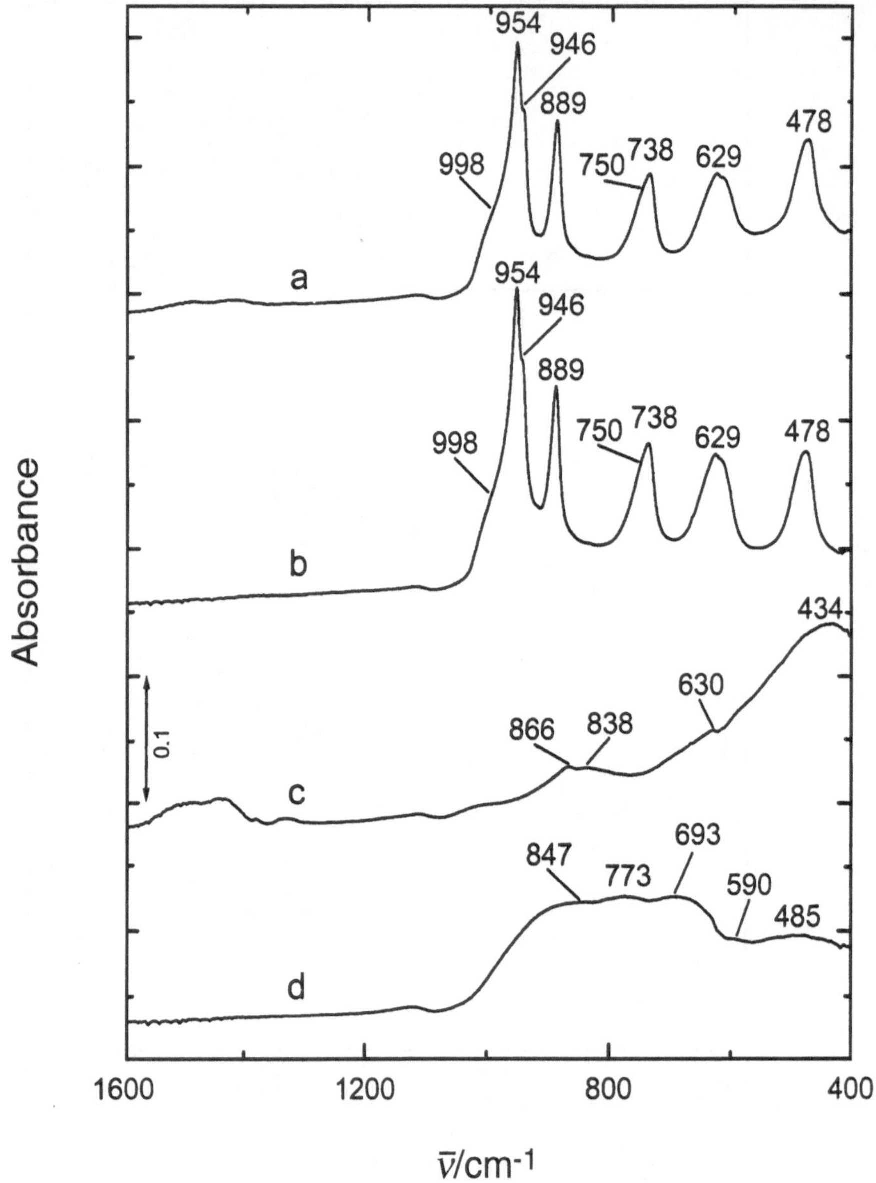

Fig. 11. *Ex situ* IR absorbance spectra of an $InVO_4$ film (500°C, 1 h): a) charged to $-0.04\ mC \cdot cm^{-2} \cdot nm^{-1}$, b) discharged from $-0.04\ mC \cdot cm^{-2} \cdot nm^{-1}$, c) charged to $-0.17\ mC \cdot cm^{-2} \cdot nm^{-1}$, d) discharged from $-0.17\ mC \cdot cm^{-2} \cdot nm^{-1}$

It is clear from curves a and b in Figs. 10–12 that small inserted charges lead to a decrease in the intensity of the IR bands, whereas their positions remain almost unaltered. In the spectra of charged films (Figs. 10–12a) an increase in the background absorption is also observed, and this is most evident for the $InVO_4$ film (Fig. 11a). This effect was ascribed to Li^+–O stretching, since these modes are expected in the spectral region below $600\ cm^{-1}$. Previous IR spectroscopic studies [31, 32] also have shown that only the tetrahedrally coordinated Li^+ ions give rise to the IR bands between 400 and $600\ cm^{-1}$. After discharging (Figs. 10–12b) we detected spectra very similar to those of the initial films; this confirmed the

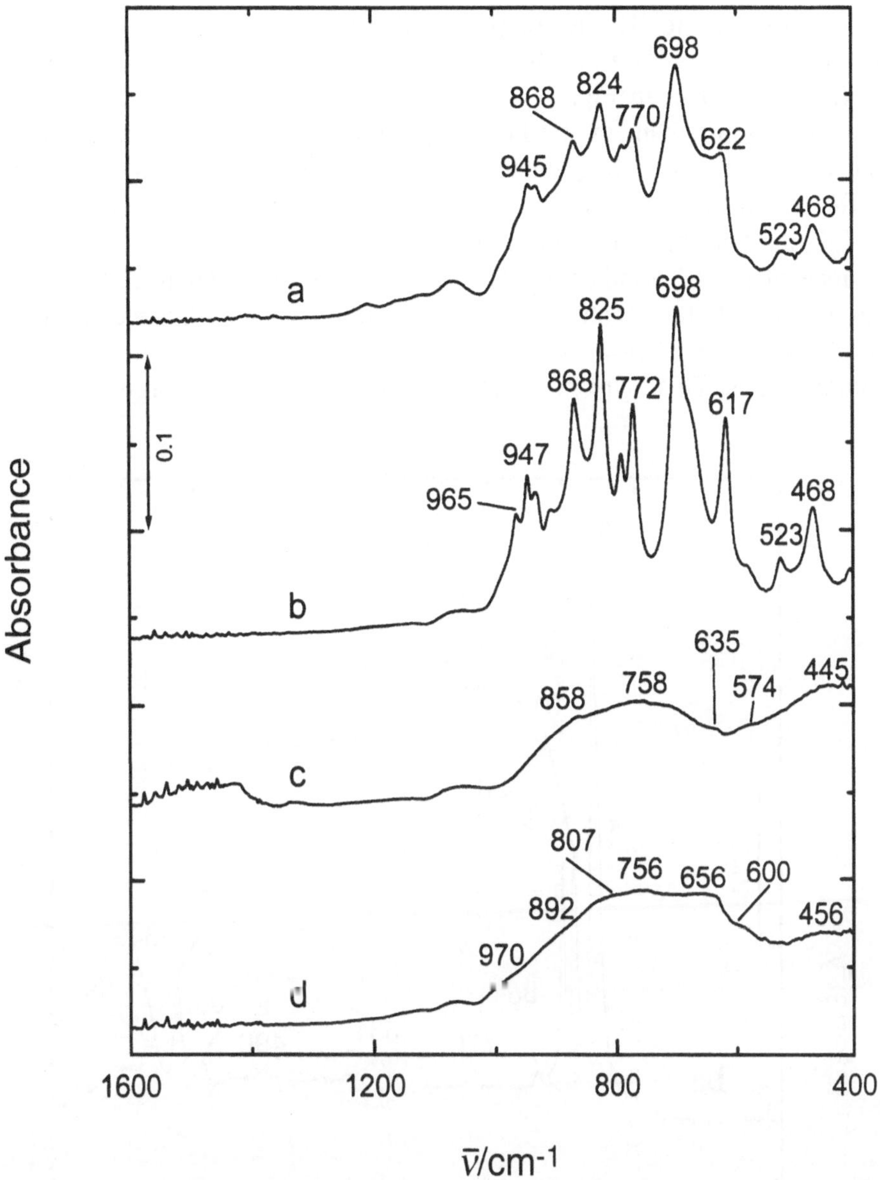

Fig. 12. *Ex situ* IR absorbance spectra of an $Fe_2V_4O_{13}$ film (400 °C, 1 h): a) charged to $-0.06\,mC \cdot cm^{-2} \cdot nm^{-1}$, b) discharged from $-0.06\,mC \cdot cm^{-2} \cdot nm^{-1}$, c) charged to $-0.33\,mC \cdot cm^{-2} \cdot nm^{-1}$, d) discharged from $-0.33\,mC \cdot cm^{-2} \cdot nm^{-}$

reversibility of the intercalation reaction. The background absorption was no longer visible.

A completely different situation was found for high chargings (Figs. 10–12c). Strong skeletal modes between 500 and 1000 cm^{-1} diminished in intensity and were substituted by broad bands with much lower intensities. On the other hand, a new and quite intense band appeared below 500 cm^{-1} and was superimposed on the background absorption. This band could be ascribed to the V–O–V (in the case of $Fe_2V_4O_{13}$) or V–O $\cdots$ M^{3+} (M = In, Fe) bridging stretching vibrations of vanadium in the reduced state (4+ or 4+/3+). Alternatively, it can be partly, together with the increase in the background absorption, attributed to the Li^+–O interactions.

Discharging led to IR spectra (Figs. 10–12d) that were quite similar to the IR spectra of the Fe–V–O and In–V–O films thermally treated at lower temperatures (300°C) shown in Figs. 7–9. This similarity is more noticeable for both orthovanadate films than for the $Fe_2V_4O_{13}$ films: the intensities of the bands are similar, whereas the band frequencies are changed to a small extent. The similarity of the IR spectra shows that the amorphization occurred in initially crystalline films after high discharging. The amorphization is, however, not complete, and the appearance of the IR bands suggets that the nanograins remained in the highly discharged films. In the spectra of discharged orthovanadate films we also observed an increased intensity below 600 cm^{-1} compared to the spectra of the initial low-temperature films. This increase in intensity showed that the charging/discharging

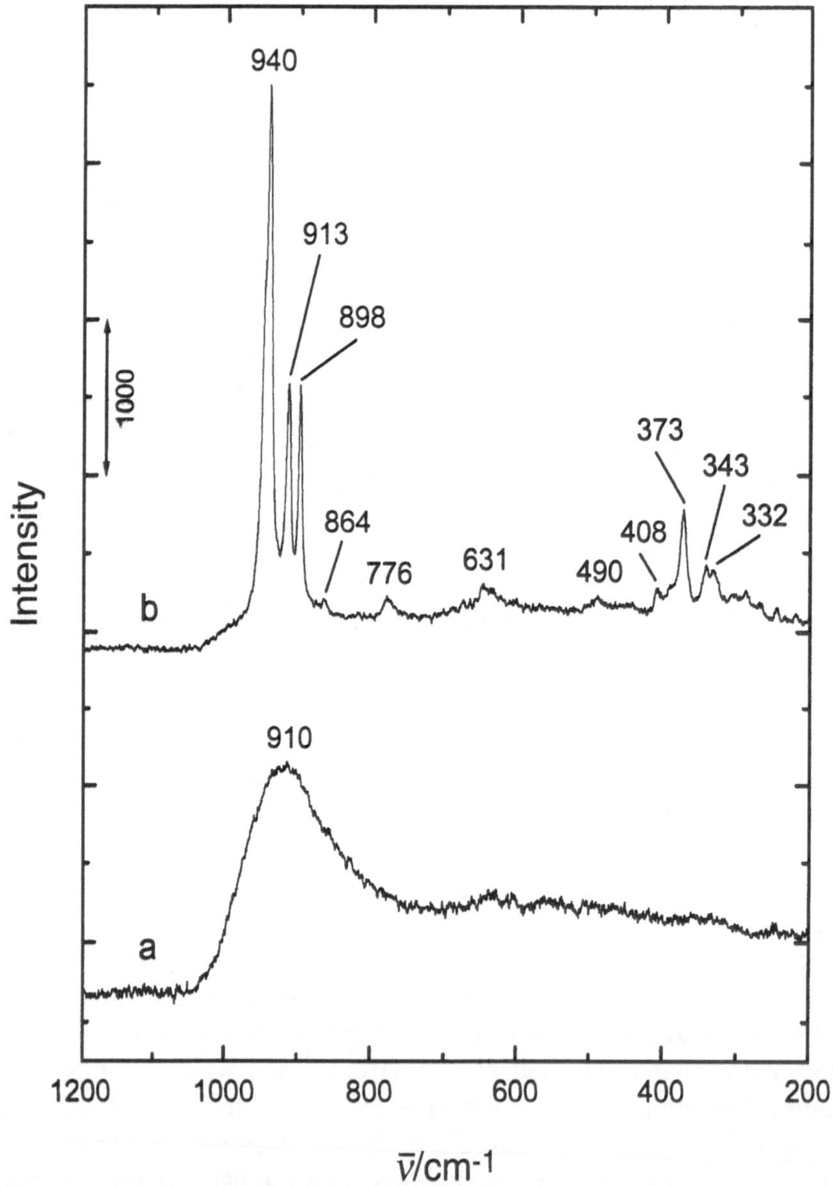

Fig. 13. *Raman* spectra of $InVO_4$ films thermally treated for 1 h at a) 300°C, b) 500°C

reaction is no longer reversible, but that the amorphisized films remain irreversibly lithiated.

Raman spectroscopy provided interesting results in the case of $InVO_4$ films. In Fig. 13, the *Raman* spectra of initial $InVO_4$ films prepared at 300 and 500°C are presented. The former spectrum has, in the skeletal region, a broad and diffuse V–O stretching band at 910 cm^{-1} (Fig. 13a). The rather large bandwidths indicate a large short-range disorder which can be due to the distribution of bond lengths and vacancies. The *Raman* spectrum of the $InVO_4$ film prepared at 500°C reveals bands at 960 (shoulder), 940, 913, and 898 cm^{-1} in the V–O stretching region. The

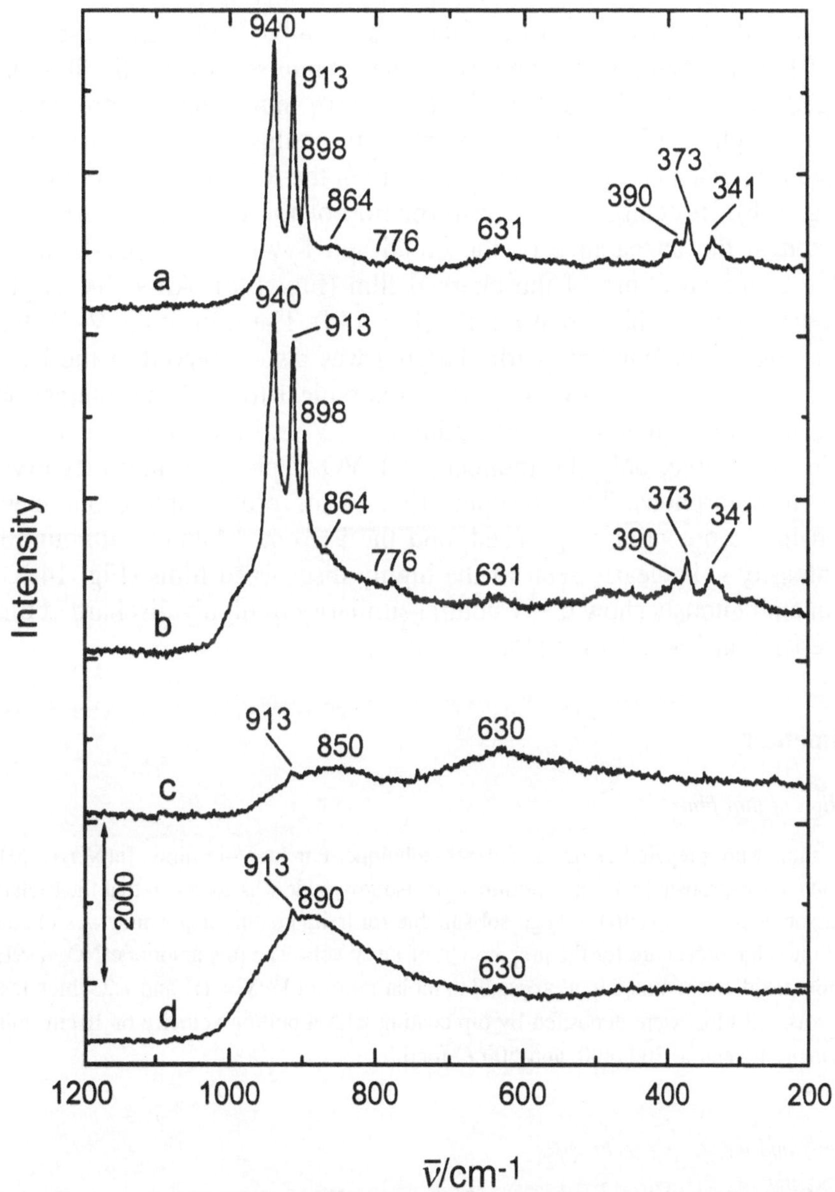

Fig. 14. *Ex situ Raman* spectra of an $InVO_4$ film (500°C, 1 h): a) charged to $-0.09\,mC \cdot cm^{-2} \cdot nm^{-1}$ ($x = 0.45$), b) discharged from $-0.09\,mC \cdot cm^{-2} \cdot nm^{-1}$, c) charged to $-0.17\,mC \cdot cm^{-2} \cdot nm^{-1}$ ($x = 0.90$), d) discharged from $-0.17\,mC \cdot cm^{-2} \cdot nm^{-1}$

$913\,cm^{-1}$ band [28] indicates the presence of the orthorhombic $InVO_4$-III phase, whereas the other bands belong to the monoclinic $InVO_4$-I phase. The exact amount of orthorhombic $InVO_4$-III is difficult to assess, but from the intensity of the $913\,cm^{-1}$ band it is certainly less than 20%.

The assignment of the $913\,cm^{-1}$ band to the orthorhombic phase was confirmed by *ex situ Raman* spectroelectrochemical measurements (Fig. 14). The $InVO_4$-I and the $InVO_4$-III phases exhibit different electrochemical behaviour [1]: the monoclinic phase becomes electrochemically active at 1.9 V *vs.* Li (−1.4 V *vs.* Ag/AgCl), whereas the orthorhombic phase requires 0.9 V *vs.* Li (−2.4 V *vs.* Ag/AgCl). Therefore, the *ex situ Raman* spectrum of the film charged to $-0.09\,mC \cdot cm^{-2} \cdot nm^{-1}$ ($x = 0.45$) revealed a drop in the intensity of the bands of the monoclinic $InVO_4$-I phase as regards the $913\,cm^{-1}$ band of the orthorhombic $InVO_4$-III phase (Fig. 14a).

The $913\,cm^{-1}$ band is also present in the *Raman* spectra of the film charged/discharged to $-0.17\,mC \cdot cm^{-1} \cdot nm^{-1}$ ($x = 0.90$), indicating the presence of the orthorhombic phase (Fig. 14a,b). The intensity of this band is, however, significantly reduced when compared to its intensity in the *Raman* spectrum of the initial film (Fig. 13b). In contrast, the transformation of the monoclinic $InVO_4$-I phase is reflected in the appearance of the broad and low-intensity bands at 850 and $630\,cm^{-1}$ in the spectrum of the charged film (Fig. 14c). After discharging, the V–O stretching mode shifts to $890\,cm^{-1}$ (Fig. 14d). The shift of the V–O stretching mode to lower wave numbers with charging was also observed in the IR spectra of charged crystalline V_2O_5 films [33] and is typical for Li^+–O interactions.

We can conclude that up to a potential of 1.7 V *vs.* Li, the range where SnO_2/F electrodes are stable, only the monoclinic $InVO_4$-I phase changes its crystalline structure to amorphous. The amorphization of the orthorhombic phase was not expected in the potential range used, and the $913\,cm^{-1}$ band – although having lower intensity – is clearly seen in the highly discharged films (Fig. 14d). These results unambiguously show the structural similarity of highly discharged films and films prepared at lower temperatures (300°C).

Experimental

Preparation of thin films

Vanadate films were prepared using the sol-gel technique. For In–V–O films, $In(NO_3)_3 \cdot 5H_2O$ was first dissolved in *n*-propanol. To this solution V-oxoisopropoxide was added so that the molar ratio in the precursors was 1:1. A yellow-orange sol suitable for a dip coating deposition was obtained. We also used the same procedure for the preparation of Fe/V sols. The precursors $Fe(NO_3)_3 \cdot 9H_2O$ and V-oxoisopropoxide were mixed in *n*-propanol in molar ratios of Fe:V = 1:1 and 1:2 which resulted in dark-red sols. All films were deposited by dip coating with a pulling velocity of $10\,cm \cdot min^{-1}$ and then thermally treated at 300, 400, and 500°C for 1 h.

Instruments and measuring techniques

To determine the film thickness we used a Profilometer Talysurf (Taylor Hobson). IR absorbance spectra of the films were measured using a Perkin Elmer System 2000 with a resolution of $4\,cm^{-1}$. For these measurements the films were deposited on double-sided polished Si wafers with an electrical resistivity of 10–20 ohm · cm and partly transparent to IR radiation (~50%). An In–Ga alloy was used to improve the electrical contact. *Ex situ* IR spectra were measured after charging/discharging of the films with either chronopotentiometry (constant current density) or chronocoulometry (constant potential) on an EG&G Par 273 potentiostat/galvanostat. In a three-electrode electrochemical cell the film was connected as a working electrode; a Pt rod was connected as a counter electrode, and a modified Ag/AgCl electrode, was used as a reference.

For *Raman* measurements, all films were deposited on SnO_2/F glass substrates. The *ex situ Raman* spectra were measured on an XY spectrograph (Dilor, France) equipped with a double monochromator as a filter and a back-illuminated liquid nitrogen-cooled 2000 × 800 pixels charge-coupled device detector (Spex, a division of the Jobin-Yvone company, France).

Samples were examined using analytical electron microscopy (AEM) in both directions: parallel (plan view) and perpendicular (cross section) to the thin film. A cross section of the Fe–V–O samples (400 and 500°C) on a ⟨Si⟩ substrate was prepared using a Gatan cross-sectional TEM specimen preparation kit. After mechanical thinning and dimpling, the samples were ion milled using 3.8 keV argon ions. To prevent degradation the sample was cooled with liquid nitrogen during the final stages of the ion erosion process. Samples were examined using a JEOL 2000 FX transmission electron microscope operating at 200 kV. The chemical composition of the phases was determined using a Link AN-1000 energy-dispersive X-ray spectroscopy (EDXS) system with an ultra-thin window Si(Li) detector. Fragments of the thin Fe–V–O film, fired at 300°C, were prepared by gentle scratching of the film's surface and subsequent transfer to a hollow carbon-coated Cu grid for AEM examination.

References

[1] Denis S, Baudrin E, Touboul M, Tarascon J-M (1997) J Electrochem Soc **144**: 4099

[2] Hayashibara M, Eguchi M, Miura T, Kishi T (1997) Solid State Ionics **98**: 119

[3] Sugawara M, Fujiwara M, Matsuki K (1993) Denki Kagaku **61**: 224

[4] Picardi G, Varsano F, Decker F, Opara Krašovec U, Šurca A, Orel B (1999) Electrochim Acta **44**: 3157

[5] Opara Krašovec U, Orel B, Reisfeld R (1998) Electrochem Solid-State Lett **1**: 104

[6] Opara Krašovec U, Orel B, Šurca A, Bukovec N, Reisfeld R (1999) Solid State Ionics **118**: 195

[7] Granqvist CG (1995) Handbook of Inorganic Electrochromic Materials. Elsevier, Amsterdam

[8] Cocciantelli JM, Doumerc JP, Pouchard M, Broussely M, Labat J (1991) J Power Sources **34**: 103

[9] Benčič S, Orel B, Šurca A, Lavrenčič Štangar U (2000) Solar Energy **68**: 499

[10] Šurca A, Orel B, Opara Krašovec U, Lavrenčič Štangar U, Dražič G (2000) J Electrochem Soc **147**: 2358

[11] Orel B, Šurca Vuk A, Opara Krašovec U, Dražič G (2001) Electrochim Acta **46**: 2059

[12] Šurca Vuk A, Opara Krašovec U, Orel B, Colomban P (2001) J Electrochem Soc **148**: H49

[13] Šurca Vuk A, Orel B, Dražič G, Decker F, Colomban P (2001) Sol-Gel Science and Technology **23**

[14] Šurca Vuk A, Orel B, Dražič G (2001) Journal of Solid State Electrochemistry **5**: 437

[15] Williams DB, Carter CB (1996) Transmission Electron Microscopy – Part II: Diffraction. Plenum, New York, p 274

[16] Rudee ML, Howie A (1972) Phil Mag **25**: 1001

[17] Graczyk JF, Chaudahari P (1973) Phys Status Solidi b **58**: 163

[18] Stadelmann PA (1987) Ultramicroscopy **21**: 131

[19] Permer L, Laligant Y (1997) Eur J Solid State Inorg Chem **34**: 41

[20] Robertson B, Kostiner E (1972) J Solid State Chem **4**: 29

[21] Oka Y, Yao T, Yamamoto N, Ueda Y, Kawasaki S, Azuma M, Tanako M (1996) J Solid State Chem **123**: 54

[22] Touboul M, Melghit K, Bénard P, Louër D (1995) J Solid State Chem **118**: 93

[23] Touboul M, Tolédano P (1980) Acta Cryst B **36**: 240

[24] Blake RL, Hessevick RE Yoltai T, Finger LW (1966) American Mineralogist **51**: 123
[25] Enjalbert R, Galy J (1986) Acta Cryst C **42**: 1467
[26] Hayakawa S, Yoko T, Sakka S (1995) J Non-Cryst Solids **183**: 73
[27] Dimitriev Y, Dimitrov V, Arnaudov M, Topalov D (1983) J Non-Cryst Solids **57**: 147
[28] Baran EJ, Escobar ME (1985) Spectrochim Acta **41**: 415
[29] Roncaglia DI, Botto IL, Baran EJ (1986) J Solid State Chemistry **62**: 11
[30] Müller A, Baran EJ, Hendra PJ (1969) Spectrochim Acta **25A**: 1654
[31] Tarte P (1962) Spectrochim Acta **18**: 467
[32] Tarte P (1964) Spectrochim Acta **20**: 238
[33] Šurca A, Orel B, Dražič G, Pihlar B (1999) J Electrochem Soc **146**: 232

Received October 4, 2001. Accepted (revised) November 23, 2001

Probing the Exciton Density of States in Semiconductor Nanocrystals Using Integrated Photoluminescence Spectroscopy

Sergey A. Filonovich[1], **Yurii P. Rakovich**[1], **Mikhail I. Vasilevskiy**[1], **Mikhail V. Artemyev**[2], **Dmitrii V. Talapin**[3], **Andrey L. Rogach**[3], **Anabela G. Rolo**[1], and **Maria J. M. Gomes**[1,*]

[1] Department of Physics, University of Minho, P-4710-057 Braga, Portugal

[2] Physico-Chemical Research Institute, Belarussian State University, 220000 Minsk, Belarus

[3] Institute of Physical Chemistry, University of Hamburg, D-20146 Hamburg, Germany

Summary. We present the results of a comparative analysis of the absorption and photoluminescence excitation (PLE) spectra *vs.* integrated photoluminescence (IPL) measured as a function of the excitation wavelength for a number of samples containing II–VI semiconductor nanocrystals (NCs) produced by different techniques. The structure of the absorption and PL spectra due to excitons confined in NCs and difficulties with the correct interpretation of the transmittance and PLE results are discussed. It is shown that, compared to the conventional PLE, the IPL intensity plotted against the excitation wavelength (IPLE spectra) reproduce better the structure of the absorption spectra. Therefore, IPLE spectroscopy can be successfully used for probing the quantized electron-hole (e-h) transitions in semiconductor nanocrystals.

Keywords. UV/Vis spectroscopy; Nanostructure; Exciton; Absorption.

Introduction

Within the last decades, the optical properties of small semiconductor particles have been intensively studied in order to obtain information on the energies and dynamics of photo-generated charge carriers as well as on the nature of the emitting states [1–4]. The electron-hole (e-h) energy levels of NCs can be determined from the analysis of the corresponding absorption spectra. However, measuring the optical absorption in films containing semiconductor nanocrystals becomes impossible if the substrate is opaque. Even if it is transparent, the collective effects in composite films with high NC fraction or *Rayleigh* scattering on film inhomogeneities often mask absorption peaks due to the quantized e-h transitions. In this work, we will demonstrate this by simple calculations and give experimental evidence of such effects in the transmittance of films containing II–VI semiconductor NCs. Also, we will show that the integrated (over a certain wavelength region) photoluminescence (PL) of NCs, measured as a function of the excitation wavelength,

* Corresponding author. E-mail: mjesus@fisica.uminho.pt

reproduces with high accuracy the shape of the absorption spectra if the latter is not obscured by other factors.

Results and Discussion

Absorption spectra modelling

Using the standard optics relations for a film on a transparent substrate, one can routinely obtain the absorption spectrum from the transmittance spectrum. The analysis of the absorption spectra can provide a way for determination of the energy spectrum of e-h pairs in NCs. Unfortunately, thin films produced by dropping NC solution onto the substrate are quite inhomogeneous in thickness (which is also hard to determine, because there are no interference fringes seen in the transmittance spectra). Unable to extract reliably the absorption, we noted that the reflectance of the films does not deviate much from unity and worked in terms of $-\log$(transmittance). Figure 1 shows experimental spectra of two *PMMA* films and one matrix-free film containing CdSe NCs in different concentration.

The modelling of the experimental spectrum of the most diluted film was performed as follows. First, the lowest absorption peak energy was taken from the spectrum. It was considered as the energy of the ($1S_e$–$1S_{3/2}$) e-h state in a

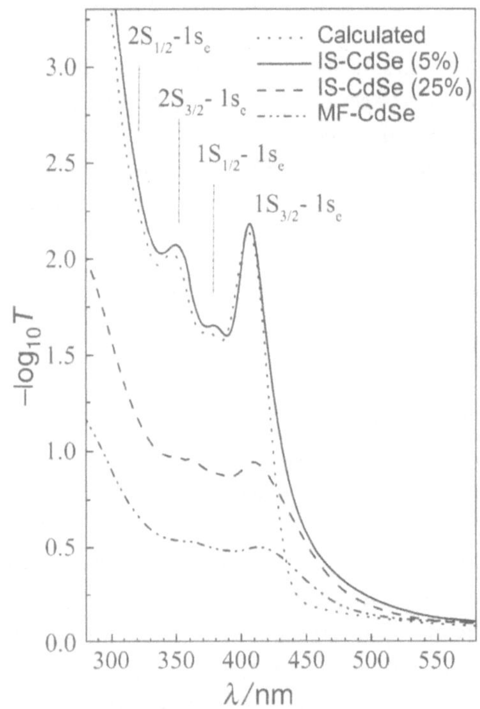

Fig. 1. Logarithm of the experimentally measured optical transmittance of two films containing CdSe NCs embedded into a *PMMA* matrices with different volume fraction and one matrix-free (MF) NC film; the theoretical curve is the optical density which was calculated assuming *Gaussian* distribution of radii (mean radius 16 Å, standard deviation 0.6 Å) and homogenous broadening of 20 meV; the film thickness was fitted

quantum dot (QD) of the most probable radius $\bar{R}$. This radius was calculated from the value of $E_1(R)$ using the effective mass approximation (EMA) [10]. Then, the energy of the $(1S_e{-}1S_{3/2})$ state for a QD of any radius R was calculated according to Eq. (1) where E_g is the bulk band gap energy, $E_{cor} = -0.248E_x$, and ε_0 and E_x are the dielectric constant and bulk exciton binding energy.

$$E_1(R) = \left(E_1(\bar{R}) - E_g - E_{cor} + 1.8\frac{e^2}{\varepsilon_0\bar{R}}\right)\left(\frac{\bar{R}}{R}\right)^2 + E_g + E_{cor} - 1.8\frac{e^2}{\varepsilon_0 R} \quad (1)$$

Secondly, the polarizability of a QD of radius R was calculated using Eq. (2).

$$A = R^3\frac{\varepsilon_\infty - \varepsilon_h + 4\pi\chi_{ef}(R)}{\varepsilon_\infty + 2\varepsilon_h} \quad (2)$$

$$\chi_{ef}(R) = \frac{3\varepsilon_h}{\varepsilon_\infty + 2\varepsilon_h}\chi(R);\ \chi(R) = \frac{e^2P^2}{m_0^2\omega^2 v}\sum_{\substack{\text{e-h}\\ \text{states}}}\frac{B_n}{E_n(R) - \hbar\omega - i\Gamma} \quad (3)$$

In Eqs. (2) and (3), ε_∞ and ε_h are the high frequency dielectric constants of the QD and matrix materials, respectively, e and m_0 are the free electron charge and mass, P is the dipole moment matrix element for the bulk semiconductor, ω is the frequency, $v = (4\pi/3)/R^3$ the QD volume, $E_n(R)$ the energy of the nth e-h state calculated according to Eq. (1), Γ is the homogeneous broadening parameter, and

$$B_n = g_n\left|\int \psi_e^{(n)}\psi_h^{(n)}d\vec{r}\right|^2 \quad (4)$$

In Eq. (4), g_n is the degeneracy factor for the nth e-h state, and $\psi_e^{(n)}$ and $\psi_h^{(n)}$ are the corresponding electron and hole EMA wavefunctions.

Given the QD polarizability A, the effective dielectric function was calculated as the third step of the spectra modelling procedure. For the film containing 5% of QDs, the modified *Maxwell-Garnett* formalism was employed which takes into account the distribution of QD radius [11]. The fitting parameters were the width of the size distribution (taken *Gaussian*), the homogeneous broadening parameter, and the film thickness. The result of these calculations presented in Fig. 1 is in quite a good agreement with the experimental spectrum.

In many cases, the absorption peaks in the spectra of films containing QDs are almost undistinguishable. A remarkable example of this is shown in Fig. 1. NCs used for preparation of all three films were synthesized in the same solution, *i.e.* were exactly the same. They differ only by NC concentration and thickness (this is why the most diluted film looks the less transparent). Why can the absorption peaks hardly be seen in the spectra of the higher NC concentration films?

One possibility is to attribute this to the dipole–dipole interaction between NCs polarized by the light. Such an effect has been observed in the FIR spectral region for dense NC films [12]. The dipole–dipole interaction can be taken into account within a well-known mean-field approximation called the *Bruggemann* model. Unfortunately, it has not been possible to include correctly the size dispersion in this scheme. Figure 2 shows what happens to the absorption when the concentration of single-size QDs increases. The peaks corresponding to the quantized e-h transition become broader and weaker and shift to the lower energy side.

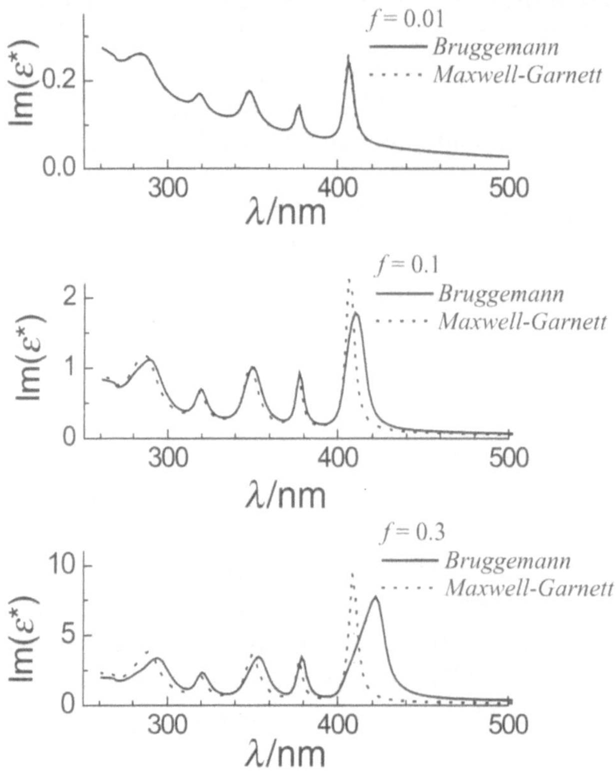

Fig. 2. Imaginary part of the effective dielectric function calculated using the modified *Bruggemann* and *Maxwell-Garnett* formalisms for ensembles of $R = 16\,\text{Å}$ CdSe QDs of different concentration; the QD parameters are the same as for the calculated spectrum in Fig. 1

Another probable reason is elastic light scattering. QDs by themselves are too small to produce significant scattering of visible light. However, they have a tendency to form aggregates in colloidal solutions and films. A simple calculation outlined below shows how the *Rayleigh* scattering can 'mask' absorption features originating from the quantized e-h transitions. First, we fitted one of the experimental spectra as explained above. Then, some *Rayleigh* scattering was assumed, which produces an extinction given by Eq. (5) (see *e.g.* Ref. [13]) where V is the volume of the inhomogeneities responsible for the scattering (for example, QD aggregates), p is the volume fraction of the inhomogeneities, and ε_1 and ε_2 are the real parts of the dielectric functions of the inhomogeneities and matrix, respectively. The former were taken as individual QDs of mean radius.

$$\gamma = \frac{4}{9} V \left(\frac{\omega}{c}\right)^4 p(1-p)(\varepsilon_1 - \varepsilon_2)^2 \tag{5}$$

The extinction according to Eq. (5) was then added to the absorption α to give an 'apparent' absorption coefficient:

$$\tilde{\alpha} = \alpha + \gamma \tag{6}$$

In Fig. 3, the calculated $\tilde{\alpha}$ spectra are shown for three values of the scattering volume. When it increases, the features in the spectrum fade out. We believe this

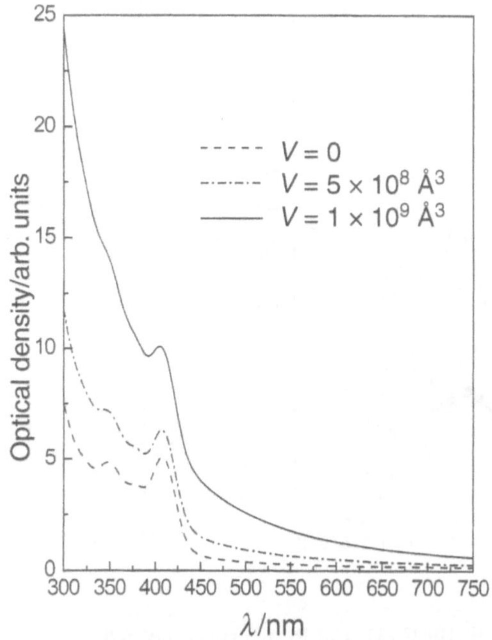

Fig. 3. Optical density spectra calculated with and without extinction (Eq. (5)); the scattering volume assumed in the calculation is given on the plot

could be the case of the higher NC concentration samples of Fig. 1, although the dipole–dipole interaction affecting real absorption was also present.

PL spectra

Let us now turn to the PL spectra shown in Fig. 4. The *Stokes* shift between the absorption and the emission for the CdSe/ZnS and the CdTe NCs was found as 31 and 104 meV, respectively. The full width of the PL band at half maximum is about 90 meV for the CdSe/ZnS nanoparticles and 155 meV for the CdTe NCs. These small shifts and widths of the PL band, which increased slightly with increasing incident photon energy, imply that the emission is due to the radiative recombination of $(1S_e{-}1S_{3/2})$ excitons in the QDs. We would like to emphasize that the PL spectra presented here were excited non-selectively, with the excitation light line intentionally broad to prevent possible size selection of the NCs. Under non-selective excitation conditions, the *Stokes* shift and PL line broadening originate mainly from the NC size distribution and fine structure of the $(1S_e{-}1S_{3/2})$ exciton state as discussed in Ref. [14]. Since the latter depends on the QD radius in a sophisticated way [14], it is rather difficult to extract the exciton density of states from the PL lineshape.

Assessing the density of states by PLE

It is well known that an absorption spectrum can be measured accurately only if the sample is sufficiently transparent, *i.e.* the condition $\alpha(\omega)d \ll 1$ must be satisfied (d is the sample thickness) [15]. As shown above, the measurement of real absorption

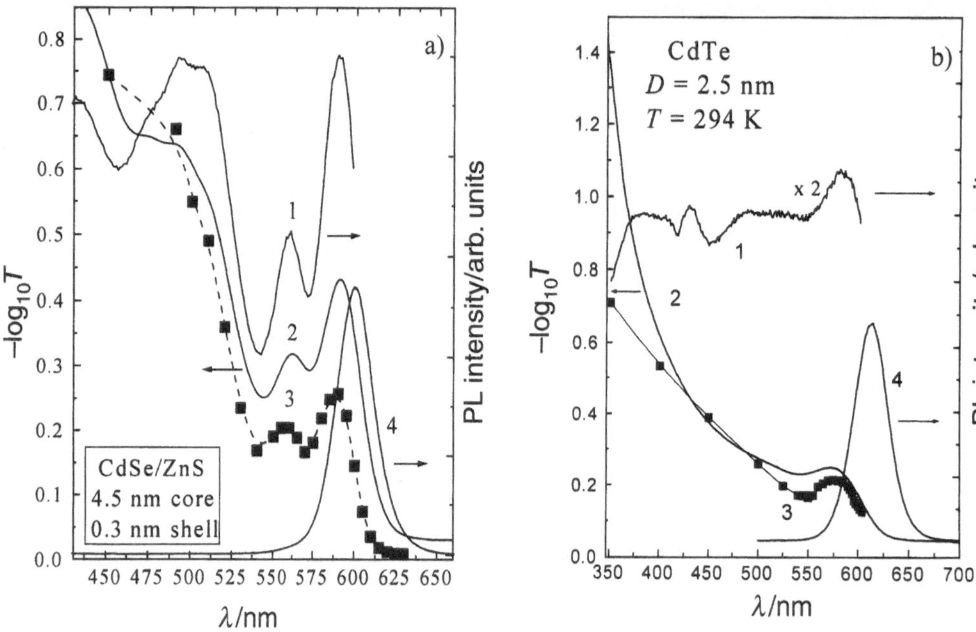

Fig. 4. PLE (1), $-\log T$ (2), IPLE (3), and PL (4) spectra of CdSe/ZnS (a) and CdTe (b) colloidal nanocrystals (T: transmittance)

can be problematic even for thin films because of the scattering and further effects. PLE spectroscopy has been generally accepted as a simple alternative in such cases. In this section, we will discuss the ability of the PLE technique for probing the density of confined excitonic states in samples containing QDs.

Measuring the PL intensity for a fixed photon energy (E_{PL}) and changing the excitation photon energy (E_{ex}), as it is usual in conventional PLE experiments [15], one should expect

$$I_{PL} \propto I_{ex} \cdot \int dR \left(g(R, E_{ex}) F(R) P_{rel} \frac{\Gamma_{PL}}{(E_1(R) - E_{PL} - \Delta(R))^2 + \Gamma_{PL}^2} \right), \qquad (7)$$

where P_{rel} is the probability of relaxation of a photogenerated e-h pair to the lowest excitonic state (E_1), $\Delta(R)$ is the *Stokes* shift between the absorption and emission in a QD of radius R, Γ_{PL} is a homogeneous broadening of the luminescence line, $F(R)$ is the QD size distribution function, and

$$g(R, E_{ex}) \propto \sum_n \frac{B_n \Gamma}{(E_n(R) - E_{ex})^2 + \Gamma^2}$$

is the excitation probability for a QD of radius R. This function can be considered independent of R for E_{ex} within the first absorption peak, because the oscillator strength does not depend on the QD radius in the strong confinement regime. However, far from the absorption edge, $g \sim R^3$, since it is proportional to the number of participating e-h states.

Assuming that the excitation process does not depend significantly on R or E_{ex} (in other words, a photoexcited e-h pair always ends up at the lowest energy state), we can put $P_{rel} = 1$. Under these conditions, PLE spectroscopy would yield a spectrum of the partial density of states due to NCs of some particular size (determined by the chosen E_{PL}) and those whose size is most close to it. Such a spectrum can hardly be correlated to the absorption of the sample.

If instead of some fixed detection energy we choose to measure the PL intensity at the (variable) wavelength corresponding to the maximum of the PL band for each E_{ex}, we should obtain

$$I_{PL}(E_{ex}) \propto I_{ex} \cdot \int g(R, E_{ex}) F(R) \frac{\Gamma_{PL}}{\Delta^2 + \Gamma_{PL}^2} dR \tag{8}$$

Although the *Stokes* shift and broadening are both size dependent [14], Eq. (8) is a better approximation to the absorption. However, the most useful approach is to integrate over the excitonic PL lineshape,

$$\int I_{PL} dE_{PL} \propto I_{ex} \cdot \int g(R, E_{ex}) F(R) dR, \tag{9}$$

and plot the integrated PL against the excitation energy. The integral in the right-hand side of Eq. (9) is just the sample absorption (in the limit of low NC concentration). Hence, such an IPLE spectrum should give a good estimate of the shape of the absorption spectrum.

IPLE spectra

Guided by the reasoning presented above, we measured PLE (detected at the maximum of the PL band) and IPLE spectra of the films containing chemically prepared NCs (see Fig. 4). Although the PLE spectra clearly correlate with the first absorption peak, they do not follow the absorption spectra for higher energies. This is because only NCs of a certain size contribute to the emission at the maximum of the PL band. In other words, the luminescence lineshape depends on the excitation wavelength as already mentioned. At the same time, the integrated PL intensity plotted *vs.* the excitation wavelength reproduces with high accuracy the absorption spectra of the NCs (Fig. 4). The good correspondence obtained between the spectral position of the peaks in the absorption and IPL spectra confirms the origin of the observed PL band as the radiative recombination of e-h pairs confined in NCs of all sizes. Thus, we propose measurements of IPLE spectra as a method alternative to absorption spectroscopy. Arguments justifying this rely on the assumption of $P_{rel} = 1$ independently of the incident photon energy. Certainly, in some cases the luminescence efficiency can depend on E_{ex}, but it apparently does not for the samples of Fig. 4.

In order to demonstrate the usefulness of IPL spectroscopy, we investigated the photoluminescence properties of a Schott OG530 cut-off filter of 3 mm thickness. The CdS_xSe_{1-x} NCs in this sample are relatively large (*ca.* 12 nm in size), and x is nominally 0.7 [16]. Figure 5 shows the absorption, PL, PLE, and IPLE spectra of the sample. The PL spectrum consists of a broad low-energy (LE) band centred at 645 nm and a pronounced high-energy (HE) peak whose intensity depends on the excitation wavelength. The IPL intensity values (squares in Fig. 5) were obtained for certain excitation wavelengths by integrating over the spectral contour of the HE peak. The LE band is not directly related to confined excitons, and therefore it was not included in the integration procedure. Because of the high thickness of the

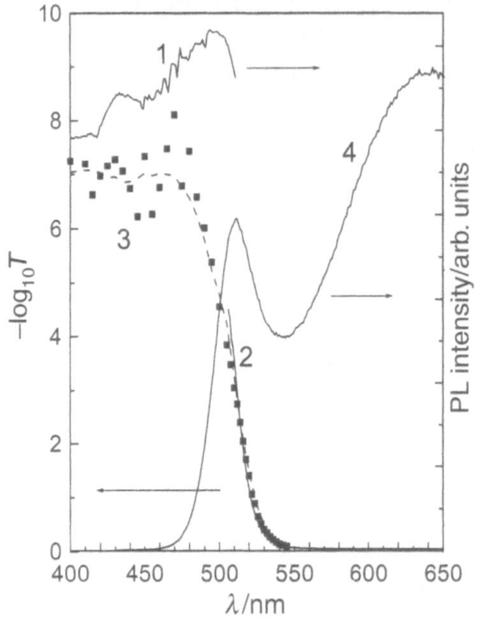

Fig. 5. PLE (1), optical density (2), IPL (3), and PL (4) spectra of OG530 glass filter; for IPL, points represent the calculated values and the line is guide to the eye

sample it is impossible to evaluate correctly the absorption from the transmission spectrum beyond 500 nm, whereas the IPLE (not the standard PLE!) allows for extrapolating the absorption spectrum up 450 nm. In the region of still shorter wavelengths, the behaviour of the IPLE spectrum becomes irregular, which manifests that the luminescence efficiency depends significantly on the excitation energy in this region. A possible reason for this can be the influence of electron or hole states (*i.e.* traps) located in the matrix or at the interface. Because of the large NC size, one should not expect noticeable confinement-related features in the absorption spectrum of this sample.

Conclusions

We measured the optical spectra of a set of samples containing II–VI nanocrystals. By modelling the absorption spectra of low NC concentration films, we were able to estimate the NC size dispersion and to evaluate the homogeneous broadening parameter. Possible reasons of smearing of the peaks in the absorption spectra are discussed. We showed that the PL intensity integrated over the excitonic PL band (rather than that measured at a fixed detection wavelength) plotted against the excitation wavelength is a function which can reproduce, under certain conditions, the absorption spectrum of a sample containing semiconductor NCs. It was demonstrated experimentally that the IPLE spectra indeed follow the absorption where the latter can be measured. It is therefore suggested that in many cases IPLE spectroscopy can be used for probing the exciton density of states in nanocrystals when direct measurements of the absorption are difficult either because of high thickness of the sample or significant extinction by film inhomogeneties.

Experimental

We studied several kinds of samples produced by different techniques. CdTe nanocrystals were prepared in the form of an aqueous colloidal solution. They were capped by a thioglycolic acid (*TGA*) shell, which made them air-stable and processable at ambient conditions. The details of the synthesis have been reported in Refs. [5, 6]. Briefly, 500 cm^3 of solution containing 0.013 *M* $Cd(ClO_4)_2$ and 0.02 *M* *TGA* were adjusted to *pH* 11.2 by addition of a 1 *M* solution of NaOH. Then, H_2Te gas obtained by decomposition of 0.8 g Al_2Te_3 with an excess of 0.05 *M* H_2SO_4 was passed into this solution in a nitrogen flow at room temperature. Subsequent heating of the reaction mixture at 100°C for different periods of time allowed to prepare colloids of CdTe nanocrystals with a mean particle size ranging from 2 to 6 nm and a particle size dispersion of about 10–15%.

CdSe and CdSe/ZnS core-shell nanocrystals were prepared as reported in Refs. [7, 8]. CdSe nanocrystals were synthesized *via* high-temperature thermolysis of Cd and Se organometallic precursors in a three-component mixture of hexadecylamine (*HDA*), tri-*n*-octylphosphine oxide (*TOPO*), and tri-*n*-octylphosphine (*TOP*). In a typical synthesis, 1 mmol of *TOP*Se, prepared according to Ref. [8], and 1.35 mmol of dimethylcadmium were dissolved in 5 cm^3 of *TOP* and rapidly injected into a vigorously stirred mixture of 10 g of *TOPO* and 5 g of *HDA* heated to 300°C. Further growth occurred at 250–310°C depending on the desirable NC size. To prepare CdSe/ZnS core-shell nanocrystals, the ZnS shell was grown onto the CdSe NCs. A solution of 0.4 mmol of diethylzinc and 0.51 mmol of bis-trimethylsilylsulfide in 3 cm^3 of *TOP* was introduced dropwise into a flask containing 2.5 cm^3 of the crude solution of CdSe NCs mixed with 5 g of *TOPO* and 2.5 g of *HDA* at 220°C.

Thin solid films of CdSe nanocrystals dispersed in a *PMMA* matrix were prepared as follows: CdSe NCs of a desired size were synthesized in *TOPO–TOP* as described above. Then, the NCs were precipitated with MeOH at room temperature and redissolved in $CHCl_3$ containing an appropriate amount of dissolved *PMMA*. Thin films of *PMMA* doped with CdSe NCs were prepared by casting from a $CHCl_3$ solution onto quartz glass and drying at room temperature [9]. Apart from the above mentioned samples we also studied some commercially available glass filters.

All transmission spectra were measured using a Shimadzu UV-3101 PC spectrometer. The PL spectra were recorded using a Spex Fluorolog spectrometer (1680-B monochromators with a dispersion of 1.70 nm/mm) equipped with a R943 Hamamatsu photomultiplier. The PL and PLE spectra were obtained by exciting the samples with a Xe lamp (output power: 0.1–2 mW depending on the wavelength; spot area: $\sim$10 mm^2).

Acknowledgements

This work was supported by the *Fundação para a Ciência e Tecnologia* (FCT, Portugal) through project PRAXIS/C/FIS/10128/1998 and SFB508 (Germany). *Y. P. Rakovich* and *S. A. Filonovich* acknowledge fellowships from PRAXIS XXI (FCT).

References

[1] Spanhel L, Haase M, Weller H, Henglein A (1987) J Am Chem Soc **109**: 5649
[2] Chestnoy N, Harris TD, Hull R, Brus LE (1986) J Phys Chem **90**: 3393
[3] O'Neil M, Marohn J, McLendon G (1990) J Phys Chem **94**: 4356
[4] Bawendi MG, Carrol PJ, Wilson W, Brus L (1992) J Chem Phys **96**: 946
[5] Rogach AL, Katsikas L, Kornowski A, Su D, Eychmüller A, Weller H (1996) Ber Bunsenges Phys Chem **100**: 1772
[6] Gao M, Kirstein S, Möhwald H, Rogach AL, Kornowski A, Eychmüller A, Weller A (1998) J Phys Chem B **102**: 8360
[7] Talapin DV, Rogach AL, Kornowski A, Haase M, Weller H (2001) Nano Lett **1**: 207
[8] Murray CB, Norris DJ, Bawendi MG (1993) J Am Chem Soc **115**: 8706

[9] Artemyev MV, Bibik AI, Gurinovich LI, Gaponenko SV, Woggon U (1999) Phys Rev B **60**: 1504
[10] Ekimov AI, Hache F, Schanneklein MC, Ricard D, Flytzanis C, Kudryavtsev IA, Yazeva TV, Rodina AV, Efros AL (1993) J Opt Soc America B **10**: 100
[11] Vasilevskiy MI, Akinkina EI, de Paula AM, Anda EV (1998) Semiconductors **32**: 11
[12] Vasilevskiy MI, Rolo AG, Artemyev MV, Filonovich SA, Gomes MJM, Rakovich YuP (2001) Physica Status Solidi B **224**: 599
[13] Landau LD, Lifshitz EM (1982) Electrodynamics of Continuous Media. Nauka, Moscow, p 582
[14] Efros AL, Rosen M, Kuno M, Nirmal M, Norris DJ, Bawendi M (1996) Phys Rev B **54**: 4843
[15] Klingshirn CF (1997) Semiconductor Optics. Springer, Berlin, p 403
[16] Persans PD, Tu AN, Lewis M, Driscoll T, Redwing R (1990) Mat Res Soc Symp Proc **164**: 105

Received October 16, 2001. Accepted (revised) January 7, 2002

EXAFS Investigations on Nanocomposites Composed of Surface-Modified Zirconium and Zirconium/Titanium Mixed Metal Oxo Clusters and Organic Polymers

Guido Kickelbick[1,*], **Martin P. Feth**[2], **Helmut Bertagnolli**[2], **Bogdan Moraru**[1], **Gregor Trimmel**[1], and **Ulrich Schubert**[1]

[1] Institut für Materialchemie, Technische Universität Wien, A-1060 Wien, Austria
[2] Institut für Physikalische Chemie, Universität Stuttgart, D-70550 Stuttgart, Germany

Summary. The surface-modified oxometallate clusters $Zr_6(OH)_4O_4(OMc)_{12}$, $Ti_4Zr_4O_6(OBu)_4(OMc)_{16}$, and $Ti_2Zr_4O_4(OBu)_2(OMc)_{14}$ (OMc = methacrylate) as well as their nanocomposites with polystyrene, poly(methacrylic acid) and poly(methyl methacrylate) were investigated by EXAFS. Studies on the nanocomposites revealed that the structure of the cluster core is retained in the hybrid materials.

Keywords. Clusters; EXAFS spectroscopy; Nanostructures; Polymerizations.

Introduction

The formation of nanocomposites embedding inorganic nanostructures in organic polymers is one of the promising methods to get new functional materials [1–5]. Phase separation is a major problem that often occurs in this kind of materials, which can be avoided by covalent bonding between the inorganic phase and the organic polymer. The attachment of organic groups with crosslinking capability onto the surface of the inorganic species allows incorporating the inorganic component by polymerization reactions. This has been proved, for example, by the surface modification of silica particles [6–11]. To expand the properties of the composite materials, new inorganic components have to be included. We have shown in earlier work that it is possible to prepare surface-modified clusters of zirconium and titanium as well as mixed zirconium–titanium systems by an *in situ* functionalization during formation of the clusters [3, 12–17]. The prepared clusters with diameters between about 0.7 and 1.7 nm of the inorganic core were incorporated in a variety of polymers, and the resulting crosslinked materials were characterized by TGA, DSC, SAXS, and NMR measurements as well as by their swelling behaviour [18, 19]. From all these methods it was not totally clear whether the cluster core was

* Corresponding author. E-mail: kickelgu@mail.zserv.tuwien.ac.at

retained in the composites. This paper reports on extended X-ray absorption fine structure (EXAFS) analyses as a powerful tool for the investigation of the local atomic environment of atoms, independent of the state of the sample [20] which was carried out on some metal oxo-clusters in organic polymers.

Results and Discussions

EXAFS investigations of the Zr_6 cluster at the Zr–K edge

The reaction of $Zr(OPr)_4$ in propanol with a 4-fold excess of methacrylic acid forms the cluster $Zr_6(OH)_4O_4(OMc)_{12}$. It consists of an octahedral $Zr_6O_4(OH)_4$ core in which the triangular faces of the Zr_6 octahedron are capped by either μ_3-O or μ_3-OH groups. Chelating and bridging methacrylate ligands saturate the remaining Zr coordination sites [12]. Inspection of the crystal structure shows that the double bonds on the surface of the cluster are fully accessible for further chemical reactions. The diameter of the cluster core of this structurally well-defined core-shell nanoparticle is approximately 0.5 nm, and that of the whole particle 1.4 nm. Nanocomposite materials were obtained by crosslinking the cluster with methyl methacrylate and methacrylic acid (50-fold molar excess of the monomer) in a free radical polymerization initiated by dibenzoyl peroxide in a benzene solution [19]. One of the major questions regarding these materials is whether the cluster was incorporated in the polymer without break-up of its structure. Results from SAXS measurements of the materials showed that the microstructure of the cluster-doped polymers can be described by a hard-sphere packing of identical spherical clusters. However, SAXS could not provide information about the molecular structure of the inorganic moiety, and therefore EXAFS investigations were carried out. The results of the EXAFS analysis are

Table 1. Structural parameters of the $Zr_6(OH)_4O_4(OMc)_{12}$ cluster (crystalline and copolymerized with methyl methacrylate and methacrylic acid) as determined from the Zr–K edge EXAFS spectrum; the coordination numbers were fixed according to the averaged crystallographic values

	[a]	$r/\text{Å}$	N	$\sigma/\text{Å}$	$\Delta E_0/\text{eV}$	k-range/Å^{-1} Fit-index
Zr_6 cluster, crystalline	Zr–O	2.09±0.02	2	0.081±0.012	21.3	3.10–16.20
	Zr–O	2.24±0.02	6	0.093±0.014		32.0
	Zr–Zr	3.52±0.04	4	0.076±0.010		
Zr_6 cluster in poly-(methyl-	Zr–O	2.10±0.02	2	0.066±0.011	20.8	3.00–16.20
methacrylate)	Zr–O	2.24±0.02	6	0.085±0.013		31.3
	Zr–Zr	3.53±0.04	4	0.075±0.010		
Zr_6 cluster in	Zr–O	2.09±0.02	2	0.071±0.010	21.5	3.00–16.20
polymethacrylate	Zr–O	2.23±0.02	6	0.084±0.013		35.2
	Zr–Zr	3.52±0.04	4	0.073±0.010		
Zr_6 cluster,	Zr–O	2.07	2			
XRD [12]	Zr–O	2.21	6			
	Zr–Zr	3.51	4			

[a] Absorber-backscatterer distance r, coordination number N, *Debye-Waller* factor σ, shift of the energy threshold energy ΔE_0, fit-index R

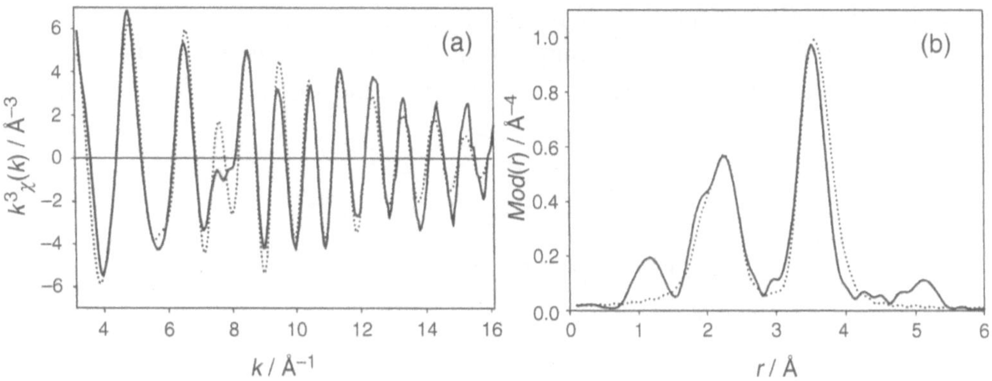

Fig. 1. Experimental (solid line) and calculated (dotted line) $k^3\chi(k)$ function (a) and their *Fourier* transforms (b) of the $Zr_6(OH)_4O_4(OMc)_{12}$ cluster at the Zr–K edge

shown in Table 1. In a first step the EXAFS data for the crystalline system were compared with the data of the structurally characterized cluster [12]. The crystal structure reveals that each Zr atom in the cluster has a first coordination sphere formed by two μ_3-O, two μ_3-OH, and four methacrylic acid O atoms. As can be seen from the *Fourier* transformed EXAFS function of the pure Zr_6 cluster (Fig. 1b), these three different oxygen types of the first coordination sphere appear as a not well-resolved double peak. The other very intense peak at a distance of about 3.5 Å can be related to zirconium backscatter. In the EXAFS analysis of the experimental k^3-weighted $\chi(k)$ function (Fig. 1a), a three-shell model can be fitted. In this analysis the coordination numbers were set to the known averaged crystallographic values [12]. The first Zr–O distance was found at 2.09 Å ($N=2$), the second at 2.24 Å ($N=6$). The Zr–Zr distance is at 3.52 Å ($N=4$). The structural parameters of the pure cluster determined by EXAFS are in very good agreement with those found from the single crystal XRD [12]. For the analysis of the nanocomposites EXAFS data, the same model for the coordination numbers was used. The hybrid materials, poly(methacrylic acid) crosslinked by the cluster, showed also a good agreement with the data in the crystalline state (Fig. 2), even the *Debye-Waller*-like factor σ remained in the same order of magnitude. This means that in both cases the original cluster structure appears to be still present in

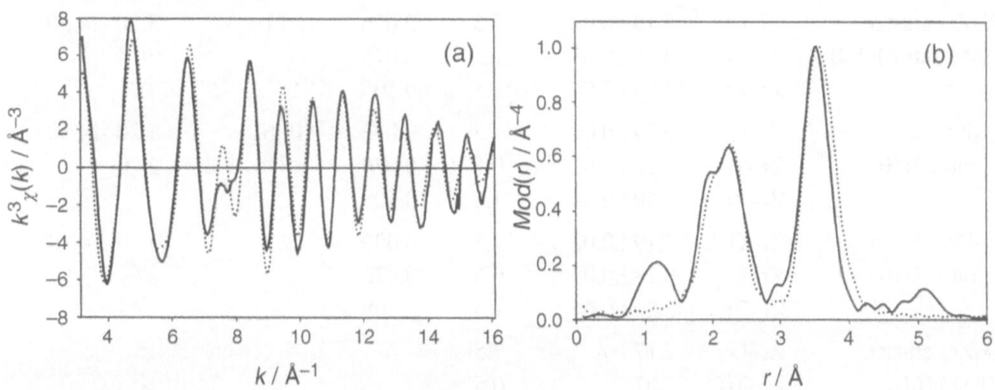

Fig. 2. Experimental (solid line) and calculated (dotted line) $k^3\chi(k)$ function (a) and their *Fourier* transforms (b) of the $Zr_6(OH)_4O_4(OMc)_{12}$ cluster in poly(methacrylic acid) at the Zr–K edge

the polymer matrix. Hence, a cleavage of the cluster by interaction with the monomers, which is a possible decomposition reaction already observed with other bidentate ligands [21] did not occur. This is a possible side reaction especially in

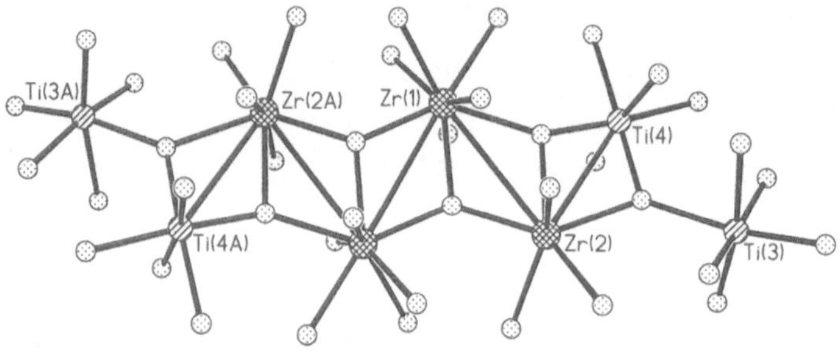

Fig. 3. Representation of the core of the $Ti_4Zr_4O_6(OBu)_4(OMc)_{16}$ cluster from X-ray structure analysis [16]

Table 2. Structural parameters of the $Ti_4Zr_4O_6(OBu)_4(OMc)_{16}$ cluster (pure as well as copolymerized with methacrylic acid (*MA*), methyl methacrylate (*MMA*), and styrene (*St*) in different ratios) as determined from the Zr–K edge EXAFS spectrum; the coordination numbers were fixed according to the averaged crystallographic values; for symbols, cf. Table 1

		r/Å	N	σ/Å	ΔE_0/eV	k-range/Å^{-1} Fit-index
Ti_4Zr_4 cluster, crystalline	Zr–O	2.20±0.02	7.5	0.086	19.9	3.20–14.30
	Zr–Ti	3.14±0.03	0.5	0.067		29.5
	Zr–Zr	3.53±0.04	1.5	0.088		
Ti_4Zr_4 cluster, *MA* ratio 1:50	Zr–O	2.19±0.02	7.5	0.093	19.9	3.10–14.20
	Zr–Ti	3.15±0.03	0.5	0.072		27.6
	Zr–Zr	3.52±0.04	1.5	0.099		
Ti_4Zr_4 cluster, *MA* ratio 1:100	Zr–O	2.20±0.02	7.5	0.094	19.9	3.10–14.20
	Zr–Ti	3.15±0.03	0.5	0.066		28.9
	Zr–Zr	3.53±0.04	1.5	0.097		
Ti_4Zr_4 cluster, *MMA* ratio 1:50	Zr–O	2.19±0.02	7.5	0.097	19.4	3.10–14.20
	Zr–Ti	3.17±0.03	0.5	0.077		27.3
	Zr–Zr	3.51±0.04	1.5	0.096		
Ti_4Zr_4 cluster, *MMA* ratio 1:100	Zr–O	2.19±0.02	7.5	0.096	20.1	3.20–14.30
	Zr–Ti	3.19±0.03	0.5	0.075		30.4
	Zr–Zr	3.50±0.04	1.5	0.105		
Ti_4Zr_4 cluster, *St* ratio 1:50	Zr–O	2.19±0.02	7.5	0.093	19.6	3.20–14.30
	Zr–Ti	3.18±0.03	0.5	0.076		27.9
	Zr–Zr	3.50±0.04	1.5	0.105		
Ti_4Zr_4 cluster, *St* ratio 1:100	Zr–O	2.19±0.02	7.5	0.093	19.7	3.20–14.30
	Zr–Ti	3.18±0.03	0.5	0.076		27.9
	Zr–Zr	3.50±0.04	1.5	0.108		
Ti_4Zr_4 cluster, XRD [16]	Zr–O	2.17	7.5			
	Zr–Ti	3.07	0.5			
	Zr–Zr	3.45	1.5			

the case of the methacrylic acid polymer with its free carboxylic acid groups. However, it cannot be excluded that surface methacrylate groups of the cluster exchange against pending groups from this polymer, which would not change the first coordination sphere around the Zr atoms.

EXAFS investigations of mixed Zr/Ti clusters at the Zr–K and Ti–K edge

Methacrylate-substituted mixed Zr/Ti clusters were synthesized by reaction of $Ti(OBu)_4$, $Zr(OBu)_4$, and methacrylic acid [16]. Depending on the Ti:Zr alkoxide ratio, different structures were obtained and characterized: $Ti_4Zr_4O_6(OBu)_4(OMc)_{16}$ at a 1:1 and $Ti_2Zr_4O_4(OBu)_2(OMc)_{14}$ at a 1:2 ratio. Both clusters show zigzag chains of $[ZrO_8]$ dodecahedra and $[TiO_6]$ octahedra building units. Contrary to the Zr_6 cluster, which can be best described as a spherical system, the metal chains in the mixed-metal compounds are arranged in an oblong shape. Particularly interesting for these compounds is the possibility to analyze the surroundings of both metal atoms due to the good X-ray absorption properties of both Zr and Ti.

Ti_4Zr_4 cluster

In $Ti_4Zr_4O_6(OBu)_4(OMc)_{16}$ the zigzag chain consists of four inner $[ZrO_8]$ dodecahedra and two $[TiO_6]$ octahedra condensed to both ends of this chain (Fig. 3).

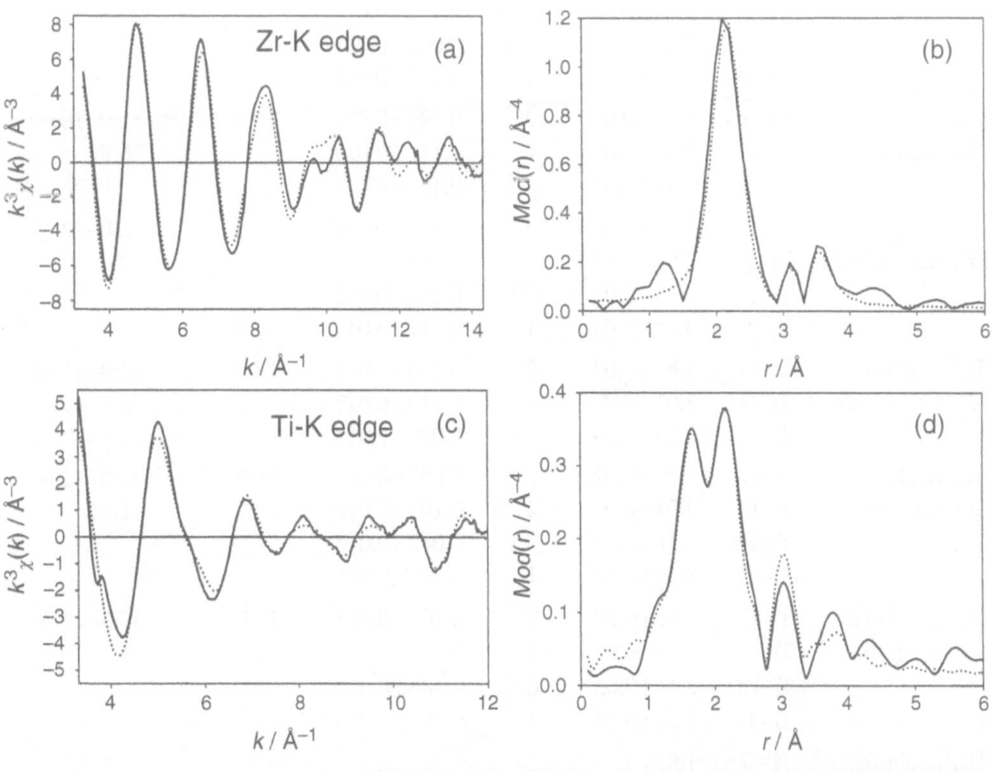

Fig. 4. Experimental (solid line) and calculated (dotted line) $k^3\chi(k)$ functions (a, c) and their *Fourier* transforms (b, d) of the pure $Ti_4Zr_4O_6(OBu)_4(OMc)_{16}$ cluster at the Zr–K (a, b) and Ti–K edge (c, d)

The average surrounding of a Zr atom in this cluster is best described by 7.5 O atoms at 2.17 Å, 0.5 Ti atoms at 3.07 Å, and 1.5 Zr atoms at 3.45 Å according to the molecular structure from single crystal X-ray diffraction experiments as can also be seen from the results of the Zr–K edge experiments (Table 2, Fig. 4). Non-integer numbers of atoms express that not every Zr atom has the same coordination sphere. The EXAFS data are in good agreement with the crystallographic data. However, compared to the Zr_6 cluster, the iterated bond lengths are somewhat longer, especially those obtained from Ti and Zr backscatter. This is a general observation in all mixed metal compounds. The origin of this small deviation is currently not clear. The results of the Zr–K edge were supported by Ti–K edge data (Table 3). The average surroundings of the Ti atoms based on the single crystal X-ray structure is best described by 2 O atoms in a distance of 1.83 Å, 4 O atoms at 2.04 Å, 0.5 Zr atoms at 3.07 Å, and 1 Ti atom at 3.45 Å. Analysis of the EXAFS

Table 3. Structural parameters of the $Ti_4Zr_4O_6(OBu)_4(OMc)_{16}$ cluster (pure as well as copolymerized with methacrylic acid (*MA*), methyl methacrylate (*MMA*), and styrene (*St*) in different ratios) as determined from the Ti–K edge EXAFS spectrum; the coordination numbers were fixed according to the averaged crystallographic values; for symbols, cf. Table 1

		$r/\text{Å}$	N	$\sigma/\text{Å}$	$\Delta E_0/\text{eV}$	k-range/Å^{-1} Fit-index
Ti_4Zr_4 cluster, crystalline	Ti–O	1.84±0.02	2	0.090±0.014	15.8	3.30–12.00
	Ti–O	2.04±0.02	4	0.106±0.016		24.3
	Ti–Zr	3.13±0.03	0.5	0.079±0.015		
	Ti–Ti	3.51±0.04	1	0.095±0.013		
Ti_4Zr_4 cluster, *MA* ratio 1:50	Ti–O	1.83±0.02	2	0.096±0.014	16.8	3.30–10.00
	Ti–O	2.03±0.02	4	0.112±0.017		26.8
	Ti–Zr	3.14±0.03	0.5	0.112±0.016		
Ti_4Zr_4 cluster, *MA* ratio 1:100	Ti–O	1.85±0.02	2	0.105±0.016	15.6	3.30–12.00
	Ti–O	2.04±0.02	4	0.118±0.018		22.4
	Ti–Zr	3.14±0.03	0.5	0.084±0.015		
	Ti–Ti	3.49±0.04	1	0.117±0.017		
Ti_4Zr_4 cluster, *MMA* ratio 1:50	Ti–O	1.82±0.02	2	0.110±0.015	17.8	3.30–11.00
	Ti–O	2.01±0.02	4	0.119±0.017		30.7
	Ti–Zr	3.13±0.03	0.5	0.090±0.015		
Ti_4Zr_4 cluster, *St* ratio 1:50	Ti–O	1.84±0.02	2	0.098±0.015	16.0	3.30–12.00
	Ti–O	2.04±0.02	4	0.108±0.016		21.3
	Ti–Zr	3.16±0.03	0.5	0.082±0.015		
	Ti–Ti	3.50±0.04	1	0.107±0.016		
Ti_4Zr_4 cluster, *St* ratio 1:100	Ti–O	1.85±0.02	2	0.105±0.015	15.8	3.30–12.00
	Ti–O	2.04±0.02	4	0.115±0.017		25.0
	Ti–Zr	3.14±0.03	0.5	0.086±0.016		
	Ti–Ti	3.50±0.04	1	0.109±0.016		
Ti_4Zr_4 cluster, XRD [16]	Ti–O	1.83	2			
	Ti–O	2.04	4			
	Ti–Zr	3.07	0.5			
	Ti–Ti	3.45	1			

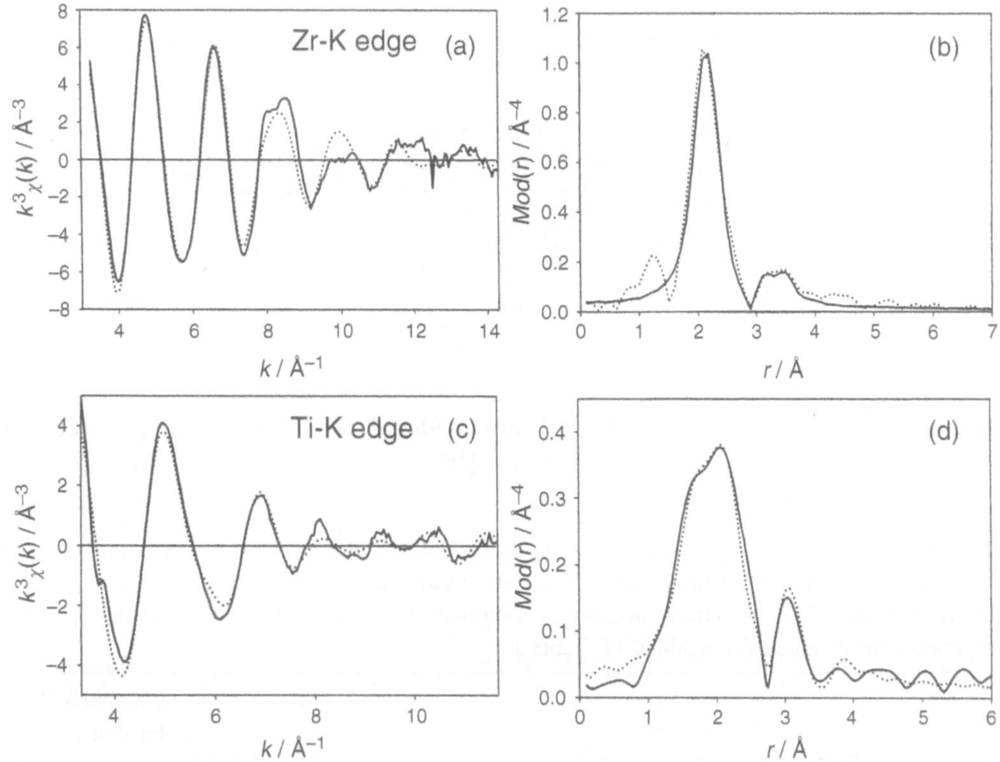

Fig. 5. Experimental (solid line) and calculated (dotted line) $k^3\chi(k)$ functions (a, c) and their *Fourier* transforms (b, d) of the $Ti_4Zr_4O_6(OBu)_4(OMc)_{16}$ cluster copolymerized with styrene (ratio 1:100) at the Zr–K (a, b) and Ti–K edge (c, d)

data from the crystals shows a good agreement with the diffraction studies. However, Zr–Ti and Ti–Ti distances were longer than in the crystal structure data as already observed for the Zr–K edge.

Nanocomposites of this cluster with poly(methacrylic acid), poly(methylmethacrylate), and polystyrene in 1:50 and 1:100 molar ratios were prepared. Altogether, the EXAFS studies reveal that in all samples the structure of the incorporated cluster is retained (Fig. 5). No influence of the type of polymer or the cluster-to-monomer ratio was observed.

The *Fourier* transforms of the Ti–K edge spectra shows that the function of the crystalline cluster exhibits a pronounced double peak (Fig. 4d), whereas the corresponding function of the copolymerized cluster has only a single, but broad peak (Fig. 5d). This qualitative result is also confirmed quantitatively. All *Debye-Waller*-like factors of the copolymerized clusters, independent of the type and ratio of the polymer, are larger than the corresponding values of the crystalline cluster, thus indicating an increase of the static disorder in the copolymerized phase.

Ti_2Zr_4 cluster

In $Ti_2Zr_4O_4(OBu)_2(OMc)_{14}$ the zigzag chain is terminated by two oxotitanium polyhedra (Fig. 6).

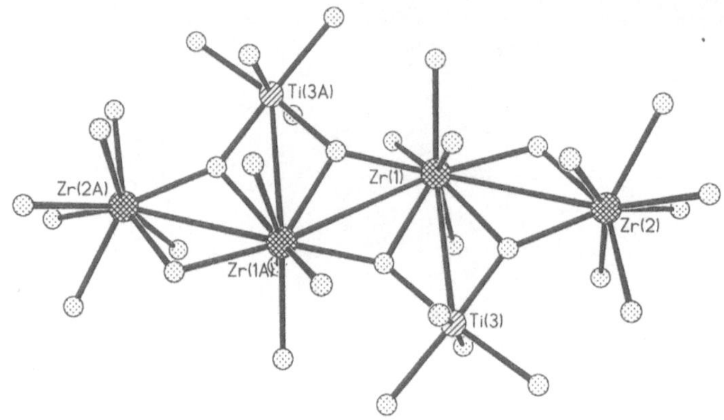

Fig. 6. Representation of the core of the $Ti_2Zr_4O_4(OBu)_2(OMc)_{14}$ cluster from X-ray structure analysis [16]

Table 4. Structural parameters of the $Ti_2Zr_4O_4(OBu)_2(OMc)_{14}$ cluster (pure as well as copolymerized with methacrylic acid (*MA*) and methylmethacrylate (*MMA*) in different ratios) as determined from the Zr–K edge EXAFS spectrum; the coordination numbers were fixed according to the averaged crystallographic values; for symbols, cf. Table 1

		r/Å	N	σ/Å	ΔE_0/eV	k-range/Å^{-1} Fit-index
Ti_2Zr_4 cluster,	Zr–O	2.20±0.02	8	0.086	20.1	3.10–14.20
crystalline	Zr–Ti	3.13±0.03	0.5	0.066		26.9
	Zr–Zr	3.54±0.04	1.5	0.073		
Ti_2Zr_4 cluster,	Zr–O	2.20±0.02	8	0.098	19.8	3.10–14.20
MA ratio 1:50	Zr–Ti	3.15±0.03	0.5	0.074		27.8
	Zr–Zr	3.52±0.04	1.5	0.094		
Ti_2Zr_4 cluster,	Zr–O	2.20±0.02	8	0.098	19.8	3.10–14.20
MA ratio 1:100	Zr–Ti	3.14±0.03	0.5	0.074		27.9
	Zr–Zr	3.52±0.04	1.5	0.094		
Ti_2Zr_4 cluster,	Zr–O	2.19±0.02	8	0.100	19.4	3.10–14.20
MMA ratio 1:50	Zr–Ti	3.18±0.03	0.5	0.073		23.3
	Zr–Zr	3.50±0.04	1.5	0.103		
Ti_2Zr_4 cluster,	Zr–O	2.20±0.02	8	0.100	19.0	3.20–14.30
MM ratio 1:100	Zr–Ti	3.17±0.03	0.5	0.077		28.2
	Zr–Zr	3.52±0.04	1.5	0.093		
Ti_2Zr_4 cluster,	Zr–O	2.20	8			
XRD [16]	Zr–Ti	3.11	0.5			
	Zr–Zr	3.51	1.5			

EXAFS analysis confirms the averaged first shell of the Zr atoms that is best described by 8 O atoms at 2.20 Å, 0.5 Ti atoms at 3.11 Å, and 1.5 Zr atoms at 3.51 Å (Table 4). The data obtained from the Zr–K edge proved, as already shown for the other nanocomposite samples, that the incorporated cluster did not structurally change in the polymer matrix. Only crystalline samples of the Ti_2Zr_4

cluster were studied at the Ti–K edge and showed good agreement with diffraction experiments (2 O atoms at 1.82 Å, 4 O atoms at 2.01 Å, and 1 Ti atom at 3.11 Å; Table 5). The comparison of the corresponding EXAFS functions (Figs. 4 and 7) additionally reveals that the environment of the metal atoms in the Ti_2Zr_4 and the Ti_4Zr_4 cluster cores are very similar.

Table 5. Structural parameters of the pure $Ti_2Zr_4O_4(OBu)_2(OMc)_{14}$ cluster as determined from the Ti–K edge EXAFS spectrum; the coordination numbers were fixed according to the averaged crystallographic values; for symbols, cf. Table 1

		r/Å	N	σ/Å	ΔE_0/eV	k-range/Å^{-1} Fit-index
Ti_2Zr_4 cluster,	Ti–O	1.83±0.02	2	0.090±0.014	17.1	3.50–13.00
crystalline	Ti–O	2.04±0.02	4	0.102±0.015		35.5
	Ti–Zr	3.13±0.03	1	0.095±0.014		
Ti_2Zr_4 cluster,	Ti–O	1.82	2			
XRD [16]	Ti–O	2.01	4			
	Ti–Zr	3.11	1			

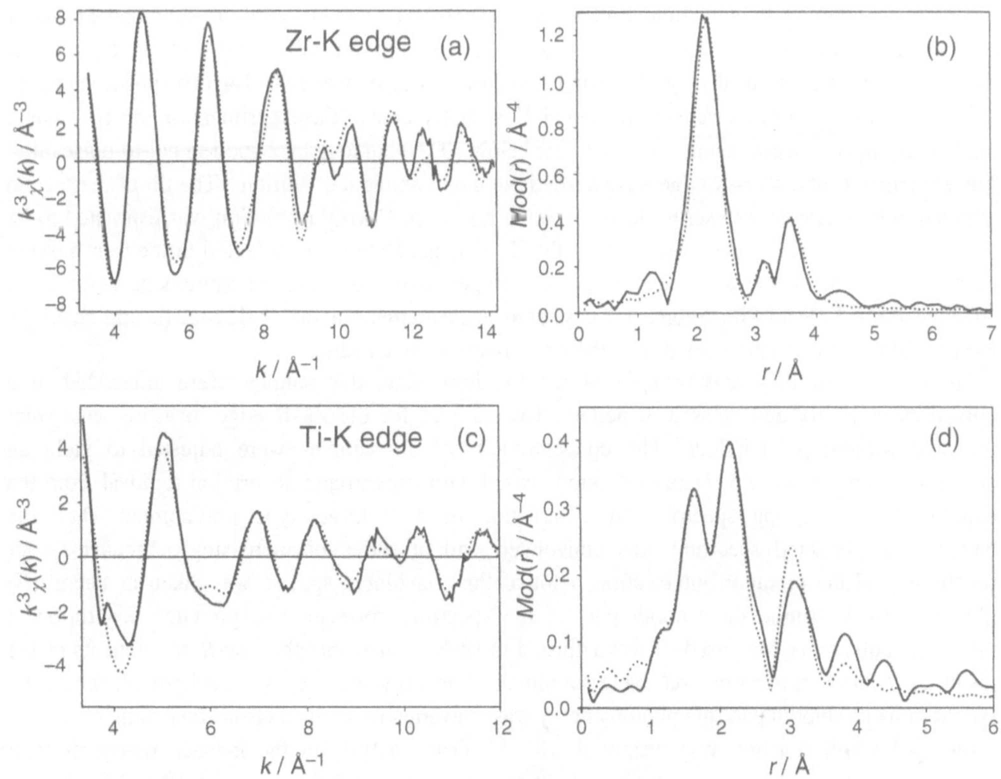

Fig. 7. Experimental (solid line) and calculated (dotted line) $k^3\chi(k)$ functions (a, c) and their *Fourier* transforms (b, d) of the pure $Ti_2Zr_4O_4(OBu)_2(OMc)_{14}$ cluster at the Zr–K (a, b) and Ti–K edge (c, d)

Conclusions

Surface-modified oxometallate clusters of Ti and Zr were investigated by EXAFS. Good agreement between crystallographic data of molecular structures and the EXAFS-analyzed systems was achieved. The analysis of nanocomposites of the clusters in different hybrid polymers reveals that the clusters are still intact in these materials, independent of the shape and composition of the cluster and also independent of the polymer type. However, in cases where the polymers can interact with the metal centers, exchange reactions of ligands with pending functionalities at the polymer chains cannot be excluded.

Experimental

Sample preparation

Surface-modified zirconium and titanium as well as mixed zirconium/titanium clusters were prepared as described in the literature [12, 16]. The nanocomposites were synthesized following a known procedure [18, 19].

EXAFS measurements and analysis

The EXAFS measurements of the samples were performed at the beamlines X1.1 (RÖMO II) and E4 at the *Hamburger Synchrotronstrahlungslabor* (HASYLAB) at DESY (Hamburg, Germany).

For the measurements at the titanium K-edge (4965.0 eV) a Si(111) double crystal monochromator and for the measurements at the zirconium K-edge (17998.0 eV) a Si(311) double crystal monochromator was used. The synchrotron beam current was between 80–140 mA (positron energy 4.45 GeV). All experiments were carried out at 25°C. In the case of the experiments at the Ti–K edge, the sample chamber was evacuated to a pressure below 10^{-4} mbar to grant a good signal-to-noise ratio. The experiments at the Zr–K edge were carried out under ambient conditions. The tilt of the second monochromator crystal was set to 30% harmonic rejection. Energy resolution was estimated to be about 1 eV for the Ti–K edge and 5 eV for the Zr–K edge. Data were collected in the transmission mode with ion chambers which were filled with nitrogen in the case of the measurements at the Ti–K edge (beamline E4) and with argon in the case of the measurements at the Zr–K edge (beamline X1.1). Energy calibration was performed with the corresponding metal foils.

In the case of the measurements at the titanium edge, the samples were embedded in a polyethylene matrix and pressed to pellets. The samples for the Zr–K edge measurements were prepared without polyethylene. The concentrations of all samples were adjusted to yield an absorption jump of $\mu \approx 1.5$. Data evaluation started with background absorption removal from the experimental absorption spectrum by subtraction of a *Victoreen*-type polynomial. Then the background-subtracted spectrum was convoluted with a series of increasingly broader *Gauss* functions, and the common intersection point of the convoluted spectra was taken as energy E_0 [22, 23]. To determine the smooth part of the spectrum corrected for pre-edge absorption, a piecewise polynomial was used. It was adjusted in such manner that the low-R components of the resulting *Fourier* transformation were minimal. After division of the background-subtracted spectrum by its smooth part, the photon energy was converted to photoelectron wave numbers k. The resulting EXAFS function was weighted with k^3. Data analysis in the k-space was performed according to the curved wave multiple scattering formalism of the program EXCURV92 with XALPHA phase and amplitude functions [24]. The mean free path of the scattered electrons was calculated from the imaginary part of the potential (VPI was set to −4.00), and an overall energy shift (ΔE_0) was assumed. The amplitude reduction factor (*AFAC*) was set to a value of 0.8 in the case of the Ti–K as well as the Zr–K edge.

Acknowledgments

We wish to thank HASYLAB at DESY, Hamburg, for the kind support of the synchrotron experiments at the beamlines E4 and A1, the *Fonds zur Förderung der wissenschaftlichen Forschung* (FWF), Austria, and the *Jubiläumsfonds der Stadt Wien* for financial support. This work was further supported by the IHP-Contract HPRI-CT-1999-00040 of the European Commission.

References

[1] Pomagailo AD (1997) Russ Chem Rev **66**: 679
[2] Pomogailo AD (2000) Russ Chem Rev **69**: 53
[3] Kickelbick G, Schubert U (2001) Monatsh Chem **132**: 13
[4] Kickelbick G, Progr Polym Sci (submitted)
[5] Schubert U (2001) Chem Mater **13**: 3487
[6] Bourgeat-Lami E, Espiard P, Guyot A, Briat S, Gauthier C, Vigier G, Perez J (1995) ACS Symp Ser **585**: 112
[7] Bourgeat-Lami E, Espiard P, Guyot A (1995) Polymer **36**: 4385
[8] Bourgeat-Lami E, Espiard P, Guyot A, Gauthier C, David L, Vigier G (1996) Angew Makromol Chem **242**: 105
[9] Bourgeat-Lami E, Lang J (1998) J Colloid Interface Sci **197**: 293
[10] Bourgeat-Lami E, Lang J (1999) J Colloid Interface Sci **210**: 281
[11] Bourgeat-Lami E, Lang J (2000) Macromol Symp **151**: 377
[12] Kickelbick G, Schubert U (1997) Chem Ber **130**: 473
[13] Kickelbick G, Schubert U (1998) Eur J Inorg Chem 159
[14] Kickelbick G, Wiede P, Schubert U (1999) Inorg Chim Acta **284**: 1
[15] Kickelbick G, Schubert U (1999) J Chem Soc, Dalton Trans 1301
[16] Moraru B, Kickelbick G, Schubert U (2001) Eur J Inorg Chem 1295
[17] Moraru B, Gross S, Kickelbick G, Trimmel G, Schubert U (2001) Monatsh Chem **132**: 993
[18] Moraru B, Hüsing N, Kickelbick G, Schubert U, Fratzl P, Peterlik H, Chem Mater (submitted)
[19] Trimmel G, Fratzl P, Schubert U (2000) Chem Mater **12**: 602
[20] Bertagnolli H, Ertel TS (1994) Angew Chem Int Ed Engl **1994**: 45
[21] Moraru B, Kickelbick G, Battistella M, Schubert U (2001) J Organomet Chem **636**: 172
[22] Ertel TS, Bertagnolli H, Hückmann S, Kolb U, Peter D (1992) Appl Spectrosc **46**: 690
[23] Newville M, Livins P, Yakoby Y, Rehr JJ, Stern EA (1993) Phys Rev B **47**: 14126
[24] Gurman SJ, Binsted N, Ross I (1986) J Phys **C 19**: 1845

Received October 23, 2001. Accepted November 12, 2001

SpringerNews

Wolfgang Linert (ed.)

Highlights in Solute-Solvent Interactions

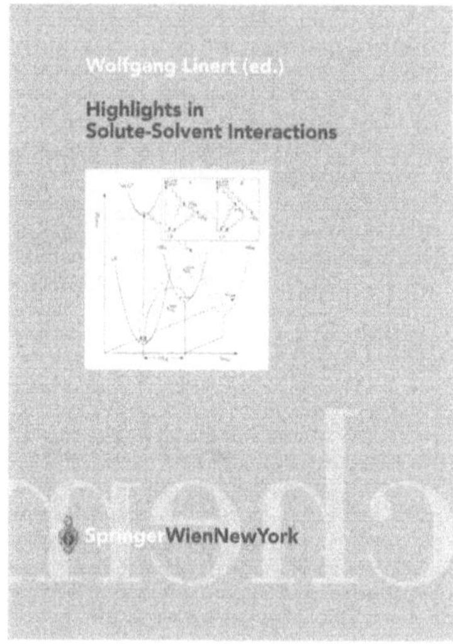

With a Foreword by Henry Taube
2002. IX, 214 pages. 87 figures.
Hardcover EUR 115,90
(Recommended retail price)
Net-price subject to local VAT.
(Special Edition of "Monatshefte für Chemie/
Chemical Monthly", Vol.132, No. 11)
ISBN 3-211-83731-0

This book emphasizes a broad spectrum of features in solution chemistry, reaching from recent developments in the empirical characterization of solvent-solute interactions up to modern theoretical descriptions of liquid-state systems, from solid-liquid interfaces to preferential solvation in mixed solvents. Accordingly, this collection presents the most important, though strongly different approaches to the understanding of the mutual influences between solute and solvent. Descriptions of actual and practically useful applications of these concepts are included.

Please visit our website: **www.springer.at**

SpringerChemistry

SpringerWienNewYork

A-1201 Wien, Sachsenplatz 4–6, P.O. Box 89, Fax +43.1.330 24 26, e-mail: books@springer.at, Internet: **www.springer.at**
D-69126 Heidelberg, Haberstraße 7, Fax +49.6221.345-229, e-mail: orders@springer.de
USA, Secaucus, NJ 07096-2485, P.O. Box 2485, Fax +1.201.348-4505, e-mail: orders@springer-ny.com
Eastern Book Service, Japan, Tokyo 113, 3–13, Hongo 3-chome, Bunkyo-ku, Fax +81.3.38 18 08 64, e-mail: orders@svt-ebs.co.jp

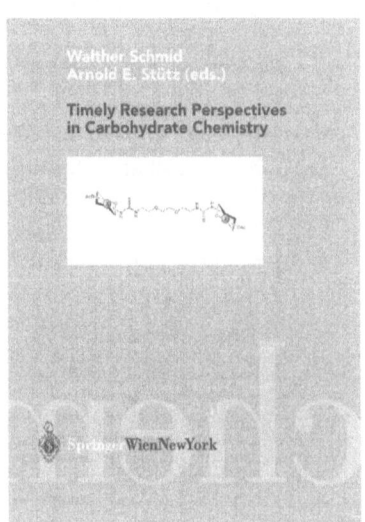